T0235843

Health, Wellbeing and Community Recovery in Fukushima

This book examines the issue of disaster recovery in relation to community wellbeing and resilience, exploring the social, political, demographic and environmental changes in the wake of the 2011 Fukushima disaster.

The contributors reflect on the Fukushima disaster of earthquake, tsunami and radiation contamination and its impacts on society from an interdisciplinary perspective of the social sciences, critical public health and the humanities. It focuses on four aspects, which form the sections of the work:

- Reflections from the Field
- Living with Risk
- Social Difference and Inequality
- Community Engagement and Wellbeing

The last three sections present research on the long-term consequences of the disaster on community health and wellbeing. These findings are enhanced and developed in the 'Notes from the Field' section, in which local practitioners from medicine and community recovery reflect on their experiences in relation to concepts developed in the previous sections.

This work significantly extends the literature on long-term wellbeing following disaster. The case study of Fukushima is a multifaceted process that illuminates wider issues around post-disaster regeneration in Fukushima. This problem takes on new importance in the context of COVID-19, including direct parallels in the issues of risk measurement, social inequality and wider wellbeing impacts from which public health disciplines can draw.

Sudeepa Abeysinghe is a health sociologist who works at the intersection of sociology, science and technology studies, and critical policy studies.

Claire Leppold is researches health and wellbeing in the aftermath of compound disasters, with a focus on the social determinants of health and health inequalities.

Akihiko Ozaki is a surgical oncologist and public health practitioner based in Fukushima, Japan.

Alison Lloyd Williams researches children's citizenship and participation in the context of disasters and disaster risk management.

Routledge Studies in Hazards, Disaster Risk and Climate Change
Series Editor: Ilan Kelman, Professor of Disasters and Health at the Institute for Risk and Disaster Reduction (IRDR) and the Institute for Global Health (IGH), University College London (UCL)

This series provides a forum for original and vibrant research. It offers contributions from each of these communities as well as innovative titles that examine the links between hazards, disasters and climate change, to bring these schools of thought closer together. This series promotes interdisciplinary scholarly work that is empirically and theoretically informed, with titles reflecting the wealth of research being undertaken in these diverse and exciting fields.

Disasters and Life in Anticipation of Slow Calamity
Perspectives from the Colombian Andes
Reidar Staupe-Delgado

The Invention of Disaster
Power and Knowledge in Discourses on Hazard and Vulnerability
JC Gaillard

Earthquakes and Volcanic Activity on Islands
History and Contemporary Perspectives from the Azores
David Chester, Angus Duncan, Rui Coutinho and Nicolau Wallenstein

Why Vulnerability Still Matters
The Politics of Disaster Risk Creation
Edited by Greg Bankoff and Dorothea Hilhorst

Health, Wellbeing and Community Recovery in Fukushima
Edited by Sudeepa Abeysinghe, Claire Leppold, Akihiko Ozaki and Alison Lloyd Williams

For more information about this series, please visit: https://www.routledge.com/Routledge-Studies-in-Hazards-Disaster-Risk-and-Climate-Change/book-series/HDC

Health, Wellbeing and Community Recovery in Fukushima

Edited by Sudeepa Abeysinghe,
Claire Leppold, Akihiko Ozaki and
Alison Lloyd Williams

Routledge
Taylor & Francis Group

LONDON AND NEW YORK

First published 2022
by Routledge
4 Park Square, Milton Park, Abingdon, Oxon OX14 4RN

and by Routledge
605 Third Avenue, New York, NY 10158

Routledge is an imprint of the Taylor & Francis Group, an informa business

British Library Cataloguing-in-Publication Data
A catalogue record for this book is available from the British Library

Library of Congress Cataloging-in-Publication Data
Names: Abeysinghe, Sudeepa, 1985- editor. | Leppold, Claire, editor. | Ozaki, Akihiko, editor. | Lloyd Williams, Alison, editor.
Title: Health, wellbeing and community recovery in Fukushima / edited by Sudeepa Abeysinghe, Claire Leppold, Akihiko Ozaki and Alison Lloyd Williams.
Description: Abingdon, Oxon ; New York, NY : Routledge, 2022. | Series: Routledge studies in hazards, disaster risk and climate change | Includes bibliographical references and index.
Identifiers: LCCN 2021059898 (print) | LCCN 2021059899 (ebook) | ISBN 9781032022734 (hbk) | ISBN 9781032022765 (pbk) | ISBN 9781003182665 (ebk)
Subjects: LCSH: Disaster relief--Japan. | Fukushima Nuclear Disaster, Japan, 2011. | Climatic changes--Japan. | Well-being--Japan.
Classification: LCC HV555.J3 H43 2022 (print) | LCC HV555.J3 (ebook) |
DDC 363.34/80952117--dc23/eng/20220127
LC record available at https://lccn.loc.gov/2021059898
LC ebook record available at https://lccn.loc.gov/2021059899

ISBN: 978-1-032-02273-4 (hbk)
ISBN: 978-1-032-02276-5 (pbk)
ISBN: 978-1-003-18266-5 (ebk)

DOI: 10.4324/9781003182665

Typeset in Times New Roman
by SPi Technologies India Pvt Ltd (Straive)

Contents

Illustrations

Figures

Tables

Acknowledgements

This edited collection would not have been possible without the collaboration and support from many people and organisations. The editors would like to thank the authors who contributed to each chapter, as well as their collaborators and research participants in Fukushima. We are especially grateful to the professional authors (Katsuka Onoda, Rika Sato, Arinobu Hori, Yasuhiro Takamura and Shigeta Mimura) who made the time to contribute to this book in the midst of their busy nursing, medical, teaching and legal careers.

The editors are also grateful for the support and insights of colleagues who have helped with particular chapters, concepts or topics. These include Maggie Mort, Emily Adrion, Sarah Hill, Kevin White, Sean Cowlishaw, Masaharu Tsubokura and other colleagues who have been so generous with their time and expertise. The editors would also like to thank Jessica Nakaya, who translated Japanese chapters into English, and Masatsugu Tanaki for assisting with the graphic design of the maps in Chapter 1. We also extend our gratitude to Akihiro Yoshikawa for helping us better understand the technical process by which the nuclear disaster occurred, as described in Chapter 1.

This work was supported by the Economic and Social Research Council (ESRC) (grant number ES/S013903/1). This funding, for the network 'Health, Risk, Disaster (HeaRD): Social Science and Post-Disaster Community Reconstruction', allowed the authors in this collection to come together through key meetings which led to the creation of this book. We are grateful to the staff at Fukushima Medical University who facilitated the first HeaRD conference in April 2019 and the Edinburgh University Research Office staff who assisted with the subsequent HeaRD meeting held in Edinburgh in July 2019. The publication of this book was also supported by the Great Britain Sasakawa Foundation for a grant (5914) which covered translation of chapters from Japanese to English.

The authors of each chapter also extend gratitude to funders who facilitated their work presented in this book. For Chapter 6, Elstow's work was assisted by an ESRC stipend as part of the NW Doctoral Training Centre and the Japanese Society for the Promotion of Science (JSPS) Summer Program 2018 (ID SP18107). For Chapter 7, Murakami's work was partially funded by a JSPS KAKENHI grant (number JP20H04354). For Chapter 10,

Lee and Uekusa's research was partially supported by funding from JSPS KAKENHI grants (JP15K01908, JP18K01185). For Chapter 11, Tanaka thanks Tak Yuasa and Yining Fang for their assistance in data collection. For Chapter 12, Lloyd Williams's and Goto's work was supported in part by a JSPS International Postdoctoral Fellowship in Japan FY2016 (Principal investigator: ALW), the Program of the Network-type Joint Usage/Research Center for Radiation Disaster Medical Science FY2018 and FY2019 (Principal investigators: AG and ALW) and the Fund for the Promotion of Joint International Research (Fostering Joint International Research [B]) from the Japan Society for the Promotion of Science Grants-in-Aid for Scientific Research (No. 19KK0060) (Principal investigator: AG). Lloyd Williams and Goto also thank all members of the research support team in Fukushima, especially Mrs Satsuki Katsumi and Mr Shigeru Nakano. For Chapter 13, De Togni's work was supported by the AHRC scheme London Arts and Humanities Partnership, and De Togni additionally thanks all Japanese collaborators and members of the Japanese judiciary system who shared legal insights.

Thank you to everyone who made this book possible. We hope that this edited collection will be a useful documentation of the ten years following the 3.11 disaster, and we look forward to seeing what the next ten years will hold for Fukushima.

Editors

Sudeepa Abeysinghe is a health sociologist who works at the intersection of sociology, science and technology studies and critical policy studies. She is a senior lecturer in global health policy at the School of Social and Political Science, University of Edinburgh. Her work focuses on the social construction and management of health risks in the context of uncertainty. This has included work on Avian Influenza, H1N1, 3.11, environmental disaster in Indonesia and COVID-19. Sudeepa is the principal investigator of the Economic Science and Research Council–funded 'Health, Risk, Disasters (HeaRD)' Network, from which these contributions have been drawn.

Claire Leppold is a research fellow at the Child and Community Wellbeing Unit at the University of Melbourne. She previously worked for five years on recovery from the 3.11 disaster in Fukushima, Japan, before moving to Australia. Her research draws on social science methods to investigate the impacts of disasters, with a particular focus on inequities, the social determinants of health and what happens when people and communities are exposed to multiple disasters.

Alison Lloyd Williams Alison Lloyd Williams holds research posts at Lancaster University and the University of Hull. She draws on creative, arts-based methods to research children's citizenship and participation. Recent work has focused on the role children can play in disaster recovery and resilience building in the UK and other parts of Europe, as well as Japan, where she has been working with schools in Fukushima Prefecture to explore and promote children's involvement in community development in the wake of the 3.11 disaster.

Akihiko Ozaki is a surgical oncologist and public health practitioner based in Fukushima, Japan. He qualified as a Doctor of Medicine at the Department of Medicine, University of Tokyo in 2010. Then, he gained a PhD in Public Health at the School of Public Health, Teikyo University in 2020. His main interest is medicine, public health and the social management of cancer and other conditions in disasters and other crises. He has more than nine years of experience in clinical practice and public health and community work in Fukushima following the 2011 triple disaster. He has also been involved in work responding to the 2015 Gorkha Earthquakes in Nepal, 2019 Typhoon Hagibis and the Covid-19 pandemic.

Contributors

Louise Elstow is a PhD researcher based at Lancaster University's sociology department. Her PhD thesis considers the making of scientific information in contamination events from a science and technology studies perspective. Over the course of two field trips, Louise spent six months in Japan working with a host of organisations and individuals who are making radiation visible in a variety of ways, trying to get under the skin of the data practices associated with radiation monitoring. Alongside academic endeavours, Louise is an emergency management specialist, her interest in nuclear emergency management stems from time working at Sellafield nuclear site on an improvement project instigated because of the Fukushima disaster. Louise holds an MSc in Risk, Crisis and Disaster Management from Leicester University and a BSc in German, Swedish and Economics from the University of Surrey.

Ben Epstein is a university assistant professor at Cambridge University's Department of Social Anthropology. He recently completed a PhD in anthropology from University College London on disaster mental health science, practice, and research after the Great East Japan Earthquake. His current research interests focus on issues of cross-cultural comparison, resilience, risk and vulnerability in the field of mental health and psychosocial support.

Aya Goto is a professor of health information and epidemiology at Fukushima Medical University Center for Integrated Science and Humanities. Goto's main research interests are the promotion of health literacy among health professionals, prevention of unintended pregnancy and parenting support. Since the Fukushima nuclear disaster in 2011, Goto has been working closely with public health nurses, helping them respond appropriately to concerns among parents of small children about elevated background radiation.

Arinobu Hori is a psychiatrist and director of Hori Mental Clinic in Minamisōma City, Fukushima Prefecture. His recent interest is focused on establishing a system to provide cognitive behavioural therapy for post-traumatic stress disorder and related personality problems in areas affected by the nuclear disaster. His clinical techniques include prolonged exposure therapy and schema therapy. Hori was born and raised in Tokyo. His early education revolved around psychoanalysis and phenomenological psychiatry. In 2012, after the disaster, he moved to Minamisōma City and helped reopen a local psychiatric hospital for three years before starting his clinic.

Sunhee Lee is an assistant professor in the Research Centre for Northeast Asian Studies at Tōhoku University in Sendai, Japan (up to March 2022) specialising in cultural anthropology. Her current research interests are the vulnerability and resilience of women and minorities after the Great East Japan Earthquake. Lee's research has particularly focused on the life stories of 'marriage-migrant' women from a gender and diversity perspective. Her key works include "Living as 'Foreign Brides': Reproductive Labour and International Arranged Marriage Migration" (*Migration Policy Review* 7, 2015); *Transporting the Age of Migration – People, Power and Community* (co-ed., Toshin-do, 2012); and *Hutkouwo torimodosu* (Iwanami Shoten, 2013).

Shunji Matsuoka is a professor at the Graduate School of Asia-Pacific Studies, the general manager of Waseda University Fukushima Hirono Future Research Center, and the director of Waseda Resilience Research Institute. He graduated from Kyoto University in 1988. He earned his PhD in Environmental Management from the Graduate School of Biosphere Sciences, Hiroshima University in 1998. He specializes in environmental economics, environmental policy, disaster and reconstruction studies and international development policy. He became a professor at Waseda University in April 2007 after serving as a professor at Hiroshima University. He was invited to the University of Malaya as a visiting professor in 1996 and American University in Washington, D.C., as a visiting researcher in 2000. He was also a visiting professor at the Arid Research Center at Tottori University, Japan, from 2009 to 2010. He conducts research around the Fukushima nuclear disaster, climate change policy in Asia, integrated solid waste management in Sri Lanka, desertification in Mongolia and social innovation at local cites in Japan.

Shigeta Mimura is a lawyer working in the field of disaster-related fatalities following the Fukushima Daiichi Nuclear Power Plant incident. Mimura graduated from Tokyo Metropolitan University Law School, Japan, in 2013 and studied law at the Legal Training and Research Institute, Supreme Court of Japan. He has worked as an attorney at the Hamadōri Law Office in Fukushima, where he was engaged in legal issues related to the nuclear incident (2014-2020). Since March 2020, Mimura has worked as a managing director and a group general counsel at IDI Infrastructures Inc., which manages an energy infrastructure-focused fund. He is also a director of the Community Foundation of Fukushima.

Michio Murakami was an associate professor at Fukushima Medical University from January 2015 to July 2021 and, since August 2021, is a specially appointed professor (full-time) at Osaka University. He specialises in risk science, which integrates risk assessment, risk management and risk communication and has been working on the social implementation of his and his colleagues' findings since the March 11 incident.

Katsuka Onoda is the director of the Nursing Department at Minamisōma Municipal General Hospital. She graduated as a registered nurse from the nursing school affiliated with National Hospital Organisation, Yamagata Hospital, in 1990. She then qualified in 1991 as a public health nurse at the Department of Health, Fukushima Prefectural General Hygiene Institute. She started working at Minamisōma Municipal General Hospital in 1992 and was a head nurse at its outpatient office at the time of the 2011 Fukushima disaster. She worked at Odaka Hospital in Minamisōma City as the director of the Nursing Department from 2017 to 2019.

Rika Sato is a vice director of the Nursing Department at Minamisōma Municipal General Hospital. She qualified as a registered nurse in 1992 at the nursing school affiliated with National Hospital Organisation, Mito Hospital. She joined Minamisōma Municipal Hospital in the same year and was working there at the time of the 2011 Fukushima disaster. She became a head nurse of the hospital's third ward in 2015 and a vice director in 2020.

Yasuhiro Takamura is a high school teacher based in Fukushima, Japan. Takamura obtained a Master of Science from the Department of Physics, Graduate School of Science, Yamagata University in 1998 and started working as a teacher at Fukushima Prefectural Futaba Shoyo High School. He then moved to Fukushima Prefectural Nihonmatsu Technical High School (1999–2003) and Fukushima Prefectural Sōma High School (2003–2012), where he engaged in the planning and operation of the 'Super Science High School' programme (2004–2009), designed by the Ministry of Education, Culture, Sports, Science and Technology and the 'Scientific Research Practice Activity Promotion Programme' (2009–2012), designed by the Japan Science and Technology Agency. Since 2012, he has been working as a teacher at Fukushima Prefectural Shinchi High School.

Mikihito Tanaka is a professor of science and media studies in the Journalism Course at the Graduate School of Political Science, Waseda University, Japan, and visiting researcher at the Department of Life Sciences Communications at the University of Wisconsin–Madison, USA. He earned his PhD in Molecular Biology from the University of Tokyo and has more than 10 years of experience as a journalist. Currently, he carries out research related to issues between science and society, mass/social media, and science journalism using both qualitative (e.g. critical discourse analysis) and quantitative (e.g. content analysis, social network analysis, natural language processing) methods. He is a founding member and research manager of the Science Media Centre of Japan. In 2011, his work with the SMCJ team was highly acclaimed for their service to the public in providing evidence-based scientific information during the Fukushima disaster and received an award from the National Institute of Science and Technology Policy.

Giulia De Togni is a research fellow at the University of Edinburgh. Her research focuses on risk, technology, health and human rights. She received an MSc in Social Anthropology from Oxford University (2015) and a PhD in Social Anthropology from University College London (2019) with dissertations focused on Fukushima. Drawing on her doctoral research, she wrote a book titled *Fall-out from Fukushima: Nuclear Evacuees Seeking Compensation and Legal Protection After the Triple Meltdown* published in the Nissan Institute/Routledge Japanese Studies Series (2021).

Shinya Uekusa is a disaster sociologist and currently a lecturer in sociology in the School of Language, Social and Political Sciences at the University of Canterbury in New Zealand. His current research interests are in the area of the sociology of disasters, resilience, social vulnerability, the sociology of language, migration and social sustainability. He is particularly interested in how socially disadvantaged communities, especially migrant, refugee and linguistic minority communities, respond to and cope with economic, social, political and environmental challenges.

1 The Reconstruction of Community and Wellbeing in Fukushima – Situating the Case within the Field

Sudeepa Abeysinghe, Akihiko Ozaki,
Claire Leppold and Alison Lloyd Williams

Disaster recovery is a pillar of disaster literature and policy. However, understandings of recovery have been traditionally rooted in practical strategies for infrastructural rebuilding, psychosocial support and efforts to prioritise health and wellbeing. More recently, the socially produced nature of individual- and community-level recovery from disasters has become widely acknowledged. The shift towards a deeper focus on the social and cultural consequences of disaster has been accompanied by a widening of the disciplines engaged in 'disaster studies' literature. This has led to an integration of broader definitions of wellbeing – which include social and community wellbeing – into the concept of recovery.

The idea of recovering from disasters is also socially constructed and contested in a way that mirrors the structures, power relations and discourses of the disaster-impacted site and those institutions that engage with it (Tierney & Oliver-Smith, 2012). In general, 'disaster recovery' is linked to the idea of rebuilding or improving the conditions of a disaster-affected community. Academic and practitioner literature tends to conceptualise recovery in three main ways: as an activity (e.g. things done after a disaster to help a community), as restoration (e.g. returning to 'how things were' before the disaster), or as change (e.g. establishing a 'new normal' or 'building back better'; Tierney & Oliver-Smith, 2012; Brady, 2019; AIDR, 2018). However, the process of 'recovery' can be partial, be unequally experienced across subpopulations (mirroring existing inequalities or producing new forms of social difference) and be temporally and spatially amorphous.

This book responds to the issue of disaster recovery in relation to the problem of health and wellbeing and, in doing so, demonstrates the complex and contested nature of literature and practice around recovery. Just over ten years on from the 3.11 disaster, it is clear that the Fukushima region has undergone significant social, political, demographic and environmental changes. The extent to which these efforts at reconstruction constitute 'recovery', what 'recovery' looks like and where 'recovery' begins and ends are examined through multiple perspectives and activities borne from this process. In reflecting on these issues, this interdisciplinary book aims to centre the focus on social aspects of health and wellbeing, reflecting on how particular (sub)communities have experienced these processes and what they say about the problem of reconstruction following a major disaster.

DOI: 10.4324/9781003182665-1

The 3.11 Disaster

The 3.11 disaster consisted of the Great East Japan Earthquake, and the resulting tsunami and nuclear disaster. The Great East Japan Earthquake occurred at 14:46 on 11 March 2011, 130 km to the east of Oshika Peninsula in Miyagi Prefecture. It was one of the largest earthquakes ever in Japanese history, with an estimated magnitude of 9.0 on the Richter scale. The resulting tsunami struck the coastal areas of eastern Japan, particularly the coastal areas of Fukushima, Miyagi and Iwate Prefectures, destroying or damaging millions of buildings and leaving 15,900 dead and 2,523 missing as of 28 February 2022 (National Police Agency, 2022).

The major earthquake and tsunami caused devastation. However, what makes 3.11 particularly unique, in terms of the psychosocial impacts and persistency of these impacts, was that it was accompanied by a major accident at the Fukushima Daiichi Nuclear Power Plant (FDNPP). This was categorized as level 7 major disaster (the maximum level) according to the International Nuclear Event Scale. The FDNPP is located on the coast between Futaba Town and Okuma Town in the Hamadōri Region of Fukushima Prefecture, and Units 1, 2 and 3 were in operation when the earthquake happened (with Units 4, 5 and 6 suspended). Figure 1.1 presents a map of Fukushima Prefecture, noting the location of FDNPP and the towns, villages and cities mentioned throughout this book.

Immediately after the earthquake, all the three units at FDNPP were automatically shut down as a safety measure, which required the continuous

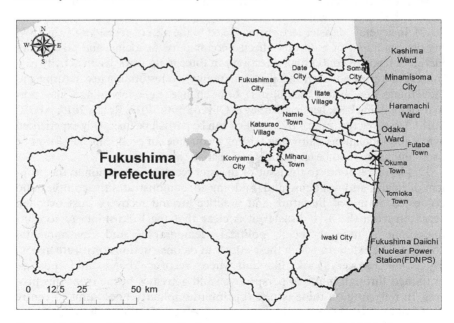

Figure 1.1 Map of Fukushima Prefecture with Fukushima Daiichi Nuclear Power Plant marked 'X'.

cooling of the reactors afterwards. However, because of the shutdown of the external power supply due to the earthquake and the inability of the machinery to cool the reactors due to the tsunami, concerns soon spread about the possibility of radioactive materials being dispersed. Then, at 20:50 on 11 March, an evacuation order was issued by the central government of Japan for the 2-km radius around the plant, and at 5:44 on 12 March 2021, the evacuation order was expanded to all areas within the 10-km radius of the power plant (Fukushima Prefectural Office, 2019; Fukushima Prefectural Office, 2020). Then, at 15:36 on 12 March 2011, the building of Unit 1 was severely damaged via a hydrogen explosion, and at 18:25 on the same day, the evacuation order was further expanded to all areas within the 20-km radius of the plant (Fukushima Prefectural Office, 2019; Fukushima Prefectural Office, 2020). This was followed by damage to the No. 3 reactor building on 14 March and to the No. 2 and No. 4 reactors on 15 March, all of which were caused by a hydrogen explosion, leading to an additional 20- to 30-km sheltering order being issued on 15 March (Fukushima Prefectural Office, 2019).

Subsequently, the evacuation zone was reconfigured on 22 April 2011, as the details of radiation exposure became clearer (Fukushima Prefectural Office, 2019). Specifically, the area within 20 km of the plant was designated as the Evacuation Order Zone (*keikai kuiki*), the area outside the 20-km zone where the annual exposure dose is likely to exceed 20 millisieverts was designated as the Planned Evacuation Zone (*keikakuteki hinan kuiki*) and the area between 20 km and 30 km from the FDNPP outside of the Planned Evacuation Zone was designated as the Emergency Evacuation Preparation Zone (*kinkyuji hinan junbi kuiki;* Fukushima Prefectural Office, 2019). As a result, for example, a predominant part of Iitate Village, which was not initially designated as an evacuation zone immediately after the accident, was additionally placed under the Planned Evacuation Zone, and residents of the village were forced to evacuate, with these evacuations lasting years. While the evacuation orders for most of Iitate Village were lifted on 31 March 2017, some former residents have continued evacuation or have chosen not to return to the village (Fukushima Prefectural Office, 2019). On the other hand, the Emergency Evacuation Preparation Zone was lifted on 30 September 2011 (Fukushima Prefectural Office, 2019), which affected the central part of Minamisōma City, enabling the city to function as the front line of the ongoing recovery work at the north edge of the evacuation zone. After 1 April 2012, when the safety of the reactors was considered to be stable, the evacuation zones were redivided into the Evacuation Order Cancellation Preparation Zone (*hinan shiji junbi kaizyo kuiki*), the Restricted Residence Zone (*kyojyū seigen kuiki*) and the Difficult-to-Return Zone (*kikan konnan kuiki*). The evacuation orders have been lifted subsequently, and all the area designated as Restricted Residence Zone and Evacuation Order Cancellation Preparation Zone became officially habitable on 10 April 2019 and 4 March 2020, respectively. As of 1 September 2021, only the Difficult-to-Return Zone remains, which covers most of Futaba Town and Okuma Town, and a part of Namie Town, leaving these areas inhabitable (Fukushima Prefectural Office,

2020). As a result of these decisions, many residents in Fukushima Prefecture were forced to evacuate, with the peak recorded number of evacuees being 62,808 displaced people on 26 January 2012 (Fukushima Prefectural Office, 2021). With a temporary transition of the evacuation zone, the number of evacuees living outside of the prefecture gradually waned, now leaving 27,998 as of 11 August 2021 (Fukushima Prefectural Office, 2021). These numbers do not include evacuation within the prefecture (i.e., internal displacement within Fukushima), meaning that many more residents in Fukushima Prefecture evacuated because of the 2011 disaster than has been captured in the official statistics.

Due to these intertwined aspects of the disaster, large areas of North-East Japan were impacted. The chapters in this book largely (although not exclusively) focus on the regions closest to the FDNPP. The movement of peoples described earlier also means that social scientific work in this area often tracks the movement of populations between places, and the impact of this on health and wellbeing. Figure 1.2 presents a map of the Tōhoku Region: the coastal areas of Fukushima, Miyagi and Iwate Prefectures were particularly affected by the disasters and are mentioned throughout this book.

In this book, when reflecting on the 3.11 disaster, we are – unless otherwise specified in individual chapters – referring to the combined consequences of the earthquake, tsunami and radiation exposure and to the disaster of the displacement of people through evacuation. This series of events and the ongoing aftermath is referred to by different nomenclature in the field. Authors writing in English from outside Japan tend towards references to 'the Fukushima disaster' or 'Fukushima triple disaster', the latter to emphasise the interlinking of three forms of hazardous event. However, for Japanese authors writing in (or translated into) English, this same series of events tends to be termed the 'Great East Japan Earthquake', which is used in Japan as a shorthand that refers to the earthquake *and* the subsequent tsunami and nuclear disaster.

In this book, Chapters 1 and 15 (i.e., the editorial team) tend towards the use of '3.11' – a term used across both Japanese and non-Japanese contexts – to refer to all aspects of the disaster and aftermath. The use of 3.11 also hopes to minimise some of the place-based stigmatisation implicit with the use of 'Fukushima' and constant co-linking of 'Fukushima' with 'disaster'. In the spirit of a multiplicity of perspectives that this volume seeks to encompass, we have tended to retain authors' original nomenclature when describing the disaster. In most cases, all terms refer to all three elements (earthquake, tsunami and nuclear incident) and their aftermath. However, as specifically identified in other cases, several authors (see, for instance, Matsuoka, Chapter 14) focus more on the impacts of the nuclear disaster, reflecting their view of this aspect as influencing the most important of the long-term impacts of 3.11.

Defining Health and Disaster

Within academic circles, there are ongoing debates about how to define *disaster* (Quarantelli, 1998; Perry, 2007) and discussions on what the focus should

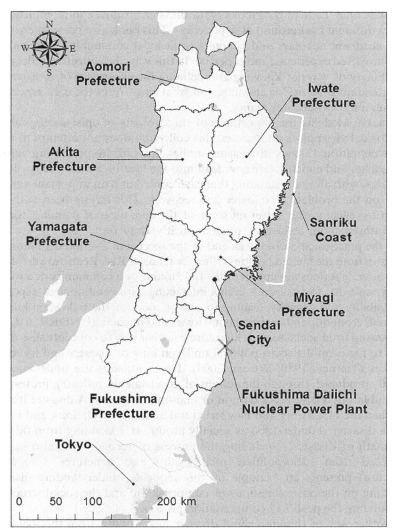

Figure 1.2 Map of the Tōhoku Region.

be when conceptualising and studying disasters. These range from a focus on curtailing ('natural' or human-made) hazards, understanding the social phenomena, measuring the scale of impacts, acknowledging underlying vulnerabilities and their role in determining disaster impacts and/or strengthening resilience and coping, among many other modes of knowing disasters. In practice, many definitional questions remain unresolved, and scholars in different disciplines often focus on different paths of inquiry.

A distinct feature of the present collection is that it brings together academics from different disciplines, as well as members of different communities in Fukushima. Each author looks at the disaster through a different lens and from a different background of experiences. This book also presents chapters from academic scholars and chapters from local community professionals who have lived experience and expertise. In this way, the collection reflects the multiplicity of ways of knowing and defining the experience of disaster and the attendant diversity of emphases on what constitutes the core aspects of disaster and post-disaster reality.

In acknowledging the importance of this diversity of epistemological and ontological viewpoints on disaster, this collection does not attempt to solve wider definitional issues in disaster studies. Instead, by presenting various approaches and constructions, we lean into the idea of a multiplicity of perspectives, with an understanding that each contributes an important vantage point on the problem of disaster and recovery. This having been said, it is useful to reflect at the outset on some of the core ways of defining disaster that authors in our collection (often implicitly) draw on.

At the practical and objectivist end of the spectrum, a sound definition can be drawn from the United Nations Office for Disaster Risk Reduction definition of *disaster*: 'a serious disruption of the functioning of a community or a society at any scale due to hazardous events interacting with conditions of exposure, vulnerability and capacity, leading to one or more of the following: human, material, economic and environmental losses and impacts' (UNDRR, n.d.).

Drawing from social science literature, another way to conceptualise disaster is to focus on the socio-political underpinning of disaster and its consequences (Tierney, 1989; Wisner, 2004). Here, disasters are understood as socially produced, through the actions of governments, industry, professions and publics in relation to a natural or man-made hazard. A disaster itself is the outcome of existing and new structural and power relations, and in this way, a disaster is understood as socially produced. Extending from this, the aftermath of disaster – including the process of recovery – is also seen as produced from socio-political relationships and structures. Chapter 5 (Mimura) presents an example of this mode of understanding disaster, reflecting on the power relations of compensation and the social structures surrounding the post-3.11 compensation process.

Another way of thinking about disaster, again drawn from the social sciences, is the idea of a 'disaster' as a socially constructed reality. Social constructionist perspectives posit that whether and how an event constitutes a disaster is a condition of the attendant discourses and ways of knowing the problem. Different groups in society can constitute the disaster, and the recovery from disaster, in different ways (Bankoff and Borrinaga, 2016; Schuller, 2016). These varied forms of knowing and experiencing have diverse consequences and can serve to socially locate the disaster in this or that place, population and period. For example, Tanaka (Chapter 11) demonstrates the power of such constructions in determining social realities, showing how the linking of disaster and radiation to the place of Fukushima Prefecture impacts individuals throughout the prefecture (not just those who were affected by the physical facts of the earthquake, tsunami and nuclear disaster).

Similarly, definitions of *health* and *wellbeing* move across these objectivist, realist and constructivist dimensions. In this volume, we take a holistic view of the concept of health. This encompasses not only physical and psychological ill health as defined by biomedical processes. It also includes the idea of social and community wellbeing, such as feelings of security and safety (see e.g. Elstow, Chapter 6) and the impacts of stigma and discrimination (see e.g. Lee, Chapter 9). Throughout the collection, we see how health can be understood as a measurable and objective reality, a product of social structures or a socially constructed reality. Chapter 3 (Hori) moves between these modes of thinking about health in its examination of the mental health consequences of 3.11. These reflections show how the incidence of mental health conditions can be a measured outcome of the disaster but simultaneously how the social location of individuals impacts their mental wellbeing and, more implicitly, how definitional categories of mental illnesses – and the status and stigma surrounding mental ill health in Japan – underpins the scale and distribution of mental illness.

More generally, this collection broadens the way in which health is understood within the disaster studies literature. The problem of health and disasters is widely studied but almost exclusively through the relatively narrow frame of 'disaster medicine'. However, health and wellbeing are not only biological and medical but also social, political and cultural products. Following disaster, health outcomes can be strongly influenced by the social environment, for example in terms of health systems impacts, how people perceive health risks, problems of social isolation following the breakdown of community or the lasting impacts of evacuation and social disruption. The chapters in this collection help to reflect on the importance of this wider understanding of 'health' following disasters, adding space for reflection on the utility of thinking about 'health' broadly when reflecting on disasters. In particular, we assert that community experiences of disaster recovery are underpinned by social understandings of health and that the field should therefore be sensitive to social framings of health and wellbeing (rather than externally imposed metrics) in its definitions of *recovery*.

Finally, in speaking to issues of recovery and resilience, the collection engages with the problem of locating a disaster in a specific time and space. Indeed, the idea of recovery itself – part of the title of this collection – implies that the disaster has now passed. Quarantelli's (1998) pivotal work 'What Is a Disaster?' indicates that a disaster is 'an event, concentrated in time and space', suggestive of the common framing of a disaster as an acute period of destablisation and crisis (the 'response' phase) to be followed by a phase of recovery. Others problematise this notion, with suggestions of the importance (and incidence) of 'slow' disasters (Erikson, 1994). This volume, while reflecting on the notion of recovery, makes no strong claims that the 3.11 disaster falls within a 'recovery' phase. Indeed, various chapters of this book (e.g. Onoda and Sato, Chapter 2) indicate that the disaster is a present reality for many individuals and communities in the area. Similarly, while the 'space' of the disaster may seem to be clearly localised within the area where the earthquake, tsunami and radiation exposure occurred, the socially constructed space of Fukushima does not neatly mirror this physical demarcation of place in relation to the felt effects or to the stigma surrounding the disaster. We are now more than ten years after the Great East Japan Earthquake, but – as the chapters in this book attest – still amidst the experience of disaster for some of the residents and former residents of this region. In this, the book makes a call for further recognition of the importance of long-term analyses of 'post'-disaster contexts, and continued reflection on reconstruction and the social impacts of this, as an addition to the importance currently placed in the field on disaster prevention, planning, and acute response.

The chapters that follow focus on different elements of the disaster; however, all look at 3.11 in terms of a serious disruption to the functioning of a community and the impacts of this disruption for different groups of people. The chapters therefore all speak to aspects to the social response and experience of post-disaster recovery, reflecting on the wide-ranging impacts on health and wellbeing that are a part of this process.

The Health Consequences of 3.11

An understanding of health consequences associated with 3.11 can be facilitated by distinguishing between aspects of the disaster and taking account of the time elapsed since the disaster. The disaster has caused 15,900 deaths, and 2,523 people remain missing as of February 2022. The primary health impacts in the immediate aftermath of 3.11 included physical injuries and drowning, which were caused by the earthquake and tsunami (National Police Agency, 2022). Since the destruction of buildings by the earthquake were limited, the tsunami was a predominant cause of these health consequences (Cabinet Office Government of Japan, 2011).

Another concern that arose immediately after the disaster and continued persistently was the health effects associated with potential radiation exposure. Looking back at history, in the aftermath of the Chernobyl nuclear disaster – the only other level 7 major nuclear accident in human history – many workers involved in the containment of the accident at the power plant died from acute radiation injury, and residents living around the plant experienced an increase of health issues, such as childhood thyroid cancer (UNSCEAR, 2010). However, after the FDNPP accident, the United Nations and United Nations Scientific Committee on the Effects of Atomic Radiation (UNSCEAR) have found that direct health effects of radiation exposure have remained at negligible levels as of the end of 2019, according to the findings of the surveys assessing internal and external exposure among the local residents (UNSCEAR, 2020). One major reason for this is that the amount of radioactive materials disseminated after the FDNPP accident was limited, only one sixth of the amount dispersed after the Chernobyl nuclear disaster (UNSCEAR, 2020). Another explanation for this is the rigorous control of food sold on the market in the aftermath of the disaster (UNSCEAR, 2020); for example, all rice produced in Fukushima Prefecture goes through inspection for radioactive materials before being sold on the market.

However, not everyone was receptive to these findings: immediately after the disaster, social networking sites were flooded with sensationalist statements about the increase in malformations in newborns and cancers caused by exposure to radiation. Stigmatisation occurred, of Fukushima as a place and of its residents, and the effects of this stigmatisation continue to this day (see also Tanaka, Chapter 11). Typical examples include the targeting of children evacuated from the prefecture for bullying and the avoidance of food produced in Fukushima. The issue of food avoidance, combined with other political conflicts, continues to this day, as illustrated by the fact that Korean athletes boycotted food from Fukushima during the Tokyo Olympics in August 2021 (*The Japan Times*, 2021).

Furthermore, it should not be overlooked that, even after the statement of the international organisations claiming limited effects of radiation exposure (UNSCEAR, 2020), there have been continuous publications in peer-review journals reporting increased cancers (see e.g. Tsuda et al., 2016). A particular problem has been childhood thyroid cancer. In the aftermath of the disaster, drawing on lessons learnt from the Chernobyl Nuclear Power Plant accident, Fukushima Prefecture and the central Japanese government continue to carry out thyroid cancer screening in children who were younger than 18 at the time of the disaster and have regularly published data from these screenings (Fukushima Medical University, 2020). The purported purpose of these efforts is to minimise uncertainty for Fukushima residents. However, some published work references an increase in thyroid cancer without reflecting on the implications of new population-wide screening in picking up cases. On the other hand, concerns about over-diagnosis have been expressed from the outset about childhood thyroid cancer screening, and the government has had to make difficult decisions about how to steer this sensitive program (Shibuya et al., 2014; Clero et al., 2021). This example demonstrates some of the unintended consequences of the policy decisions surrounding disaster events.

In addition to the preceding discussion, the existence of secondary health issues should not be overlooked. Secondary health issues are defined here as those which are assumed to have been caused by the disaster but which are not directly related to radiation exposure, or direct physical damage such as injuries and drowning. For example, in Minamisōma City and Sōma City, it has been demonstrated that many residents died from such secondary health issues in the first month after the disaster in a rise of excess deaths, with pneumonia the most common type of secondary health issue at this time (Morita et al., 2017). The reason for this phenomenon may be that immediately after the disaster, the residents of this area were not only exposed to physical and mental stress due to evacuation but also did not receive adequate care due to the collapse of the medical system. In fact, it has been pointed out that in this period, the number of doctors, nurses and other staff at medical institutions in the Hamadōri Region was reduced and that emergency transportation (during evacuation) may not have been carried out safely (Kodama et al., 2014; Morita et al., 2016).

In addition, the impact of the evacuation is extremely important in considering secondary health issues in Fukushima Prefecture. Patients who were evacuated from hospitals and long-term care facilities were a group who experienced some of the most hardship in the immediate aftermath of the disaster. For example, many residents of long-term care facilities in Minamisōma were forced to evacuate after the disaster, and there were high mortality rates during or soon after evacuation (Nomura et al., 2016a; Ohba et al., 2021). There are multiple possible reasons for this. First, many of these residents were bedridden and in need of ongoing care. Experiencing prolonged evacuation by bus or other means without adequate care may have placed undue strain on their bodies. The most extreme example of such an event was at Futaba Hospital, located 4.5 km from the nuclear power plant, where a combination of adverse conditions resulted in the deaths of about 50 people during bus evacuations between 12 and 16 March 2011. Another cause was inadequate information transfer between institutions. For many patients, in the confusion after the disaster, their information and records were not passed on sufficiently to the in-taking institution, and they may not have received the same care as before the evacuation (e.g. see Sonoda et al., 2019).

Multiple studies have since found that evacuation led to health problems not only in the immediate aftermath of the disaster but also in the medium and long term. Among those who were forced to evacuate, increased risk of non-communicable diseases (such as diabetes) and a general decline in physical function have been reported as a result of living in 'temporary' housing (for many months or years; Yabuki et al., 2015; Nomura et al., 2016b). Widespread social disruption and general deterioration of living conditions have also been related back to health risks (Hasegawa et al., 2015; Leppold et al., 2016). Evacuation has also led to a decline in the population in local towns, an ageing population and a weakening of communities. These long-term changes highlight the social aspects to disaster impacts, and the deep links between how community changes (e.g. evacuation, disrupted living conditions, pressure on the medical system) can influence health and wellbeing. This book sits directly at this intersection of the social impacts and health impacts of this disaster.

One lesson we can learn from the 3.11 disaster is that the socio-political consequences of crises are often long-lasting and wide-reaching. Notable health impacts of 3.11 have included acute problems such as injury or loss of care. Long-term health impacts include increases in chronic disease (Nomura et al., 2016b), widening of health inequalities (Aida et al., 2013) and persistent stressors to mental health (Norris et al., 2002; Wind and Komproe, 2012). Disasters also fundamentally reconfigure the structural relationships of health, including health systems effects, such the destabilisation of medical workers or long-term medical needs (Gill, 2007; Abeysinghe et al., 2017). Health and illness can become a particular public and political focal point during crisis, as is evident in 3.11, for which the potential health consequences of nuclear radiation remain a site of dispute (see Elstow, Chapter 6).

There is a dearth of research engaging with health and disasters through social science perspectives. As our overview of the evidence base above indicates, this topic therefore tends to be approached from biomedical, epidemiological or hospital management fields (see also Smith et al., 2012). Links between health and disaster within the social sciences are marginal (cf. two key reviews of disaster sociology and the dearth of attention to health: Tierney, 2007; Guggenheim, 2014). Research in this field has, as noted by a leading disaster sociologist 'remained remarkably resistant to changes in the broader [disciplinary] landscape, [where] its strong applied focus has been a barrier to theoretical innovation' (Tierney, 2007: 503); this is particularly true of health-related disaster research. Work is often drawn from quick-response case studies, and more conceptually driven research is situated within isolated sub-areas (in particular, mental wellbeing and the concept of resilience).

The 3.11 disaster continues to influence health and wellbeing, and an inter-disciplinary approach to understanding these issues is necessary. This volume adds to the emerging more interdisciplinary literature in health and disasters, highlighting social science and critical public health perspectives as an anti-dote to the dominance of more biomedical accounts within this field.

Reflections on the 3.11 Disaster

In addition to bringing together the viewpoints of different disciplines, we also present these in parallel with voices of lay and professional expertise surrounding 3.11. There is a growing comprehension that disaster-affected communities themselves are an important site of experiential expertise, with the possibility for learning through accessing these understandings (Fortun and Morgan, 2016). Here, we combine academic voices with those inhabiting professional expertise and personal experience in Fukushima post-3.11.

The book is arranged in relation to four themes: 'Reflections from the Field', 'Living with Risk and Uncertainty', 'Social Difference and Inequality', and 'Community Action, Engagement and Wellbeing'.

In highlighting the importance of non-academic forms of expertise, the first part of the book, 'Reflections from the Field', provides viewpoints from professionals engaging with post-disaster care and reflection. Chapter 2 presents a set of personal and professional reflections from nurses, Onoda and Sato, who capture the professional difficulties and motivations of working in a (post-)disaster setting. In Chapter 3, Hori provides a reflection on the inci-dence and definition of mental health disorders following the disaster (link-ing with the academic discussions in Chapter 8, Epstein). Takamura (Chapter 4) provides insights as a high school teacher, reflecting on the learning and development that his students have accomplished in light of the disaster. This connects with the chapter by Lloyd Williams and Goto in showing how work-ing with young people can be a positive way for communities to reflect, rebuild and look to the future. Finally, in Chapter 5 Mimura shows how legal processes are experienced from the practitioner perspective, linking with De Togni's work (Chapter 13). Mimura gives us an understanding of the com-plexities faced by individuals and families in engaging with the compensation process for harms incurred during the acute phases of 3.11. Linking with the theme of community knowledges and resilience (Part IV), 'Reflections from the Field' intertwines professional expertise and personal experience in pro-viding embedded accounts of living and working in the context of disaster.

Risk and a sense of the unknown or unknowability have been persistent markers of 3.11, particularly owing to the issue of nuclear contamination. This is explored in 'Living with Risk and Uncertainty'. Chapters 6, 7 and 8 reflect on different ways in which communities have attempted to negotiate and manage the risks that they face. Elstow (Chapter 6) assesses the various community practices of risk monitoring that take place in Fukushima Prefecture and demonstrates how practices of measurement and monitoring are linked to wider politics of safety, community and compensation. In Chapter 7, Murakami engages with these processes of risk measurement head-on, exploring debates on the scientific and political use of risk measures for recovery policy and proposing an indicator for capturing the concept of social health for use within policy discussions. The problem of measurement is also critically assessed in Chapter 8, where Epstein reflects on regimes of therapeutic governance surrounding mental ill health and post-disaster attempts to capture an understanding of mental wellbeing. Taken together, this section illuminates the social experience of risk and unpacks attempts by formal and informal groups of actors to create both scientifically and socially robust measurements of risk and metrics of (un)acceptable levels of risk.

Part III, 'Social Difference and Inequality', looks at the problems of marginalisation and inequality that have become amplified through the experience of the disaster. Lee focuses on the issue of gender and its intersection with minority identities, in an area of work that is under-examined in the post-3.11 literature. Here (Chapter 9), she reflects on how these intersectional identities link to the experienced impact of the disaster, advocating for the importance of gender perspectives in future disaster prevention and recovery efforts. In Chapter 10, Lee and Uekusa explore the concept of social vulnerability after the 3.11 disaster. Specifically, they look at how migrant women – situated in a context of wider inequalities in the Tōhoku Region – faced uneven impacts of the disaster. Through examining circumstances of not only invisibility, isolation and oppression but also empowerment and resilience, these authors highlight a need for migration and feminist scholars to re-examine what vulnerability and empowerment involve in disaster settings. Moving from the level of specific groups and populations, Tanaka (Chapter 11) takes a macro-sociological lens to the problem of place and disaster. In this chapter, he demonstrates the social construction of 'Fukushima' in the media and reflects on the siting of stigma. Tanaka makes a persuasive argument that the social construction of Fukushima and its people has superseded the local realities in Fukushima Prefecture in the national imagination of the disaster and its location. This section therefore takes a wide-ranging look at the way in which disaster can magnify existing inequalities (Chapter 9), intersect with lived realities (Chapter 10) and produce new forms or stigmatisation and social difference (Chapter 11).

Building on the ideas of resilience developed in Part III, 'Community Action, Engagement and Wellbeing', engages closely with ideas of empowerment and wellbeing in the context of 3.11. In Chapter 12, Lloyd Williams and Goto bring together the disciplines of public health and participatory arts to demonstrate the usefulness of creative practices within education as a means of building children's and, by extension, community resilience. De Togni (Chapter 13) then looks at community-based resilience strategies by examining modes of empowerment and powerlessness in relation to accessing the legal system in relation to the nuclear disaster. She demonstrates how the identities of individuals seeking legal recourse are transformed through new networks of trust and support via their legal challenges. Matsuoka (Chapter 14) looks at the topic of lay and experiential expertise from the other side of the policy puzzle, reflecting on ways to feed socially robust knowledge into the science–policy interface. This chapter reflects on the importance of inserting participatory and deliberative democratic processes in the post-3.11 reconstruction process. This section as a whole adds to discussions about the value of experiential and lay expertise, arguing for more to be done to promote and develop these different forms of expertise and integrate them into the post-disaster processes of resilience building.

This collection is unique in providing an approach to interrogating disasters that moves beyond quick-response studies and the acute event to an analysis of the longer-term impacts of a crisis on communities, harnesses an interdisciplinary mix of work to reflect on health and wellbeing beyond the narrow disaster medicine literature and provides space for non-academic actors to reflect on their professional experience, enriching both the presented research and providing experiential insights to the long-term post-disaster experience. In compiling this work, we hope that we can help enhance scholarly and practical understandings of the social experience of health and disaster recovery and add to ongoing debates around how to best prepare for and react to disasters.

References

Abeysinghe, S., Leppold, C., Ozaki, A., Morita, M., & Tsubokura, M. (2017). Disappearing everyday materials: the displacement of medical resources following disaster in Fukushima, Japan. *Social Science & Medicine*, *191*, 117–124.

Aida, J., Kawachi, I., Subramanian, S., & Kondo, K. (2013). Disaster, social capital, and health. In Kawachi, I., Takao, S., & Subramanian, S. V. (Eds.), *Global perspectives on social capital and health* (pp. 167–187). New York, NY: Springer.

Australian Institute for Disaster Resilience (AIDR) (2018). *Community recovery handbook* [online]. Available at: https://knowledge.aidr.org.au/resources/handbook-community-recovery/ (Accessed: 13 August, 2021)

Bankoff, G., & Borrinaga, G. E. (2016). Whethering the storm: the twin natures of typhoons Haiyan and Yolanda. In Button, G. V., & Schuller, M. (Eds), *Contextualizing disaster* (pp. 44–65). New York, NY: Berghan.

Brady, K. (2019). What do people who have been affected by a disaster consider to be helpful and unhelpful in their recovery? PhD thesis, University of Melbourne.

Cabinet Office Government of Japan (2011). *White paper on disaster management 2011* (online). Available at: http://www.bousai.go.jp/kaigirep/hakusho/pdf/WPDM2011_Summary.pdf (Accessed: 2 September, 2021)

Clero, E., Ostroumova, E., Demoury, C., Grosche, B., Kesminiene, A., Liutsko, L., Motreff, Y., Oughton, D., Pirard, P., Rogel, A., & Van Nieuwenhuyse, A. (2021). Lessons learned from Chernobyl and Fukushima on thyroid cancer screening and recommendations in case of a future nuclear accident. *Enviroent International, 146,* 106230.

Erikson, K. (1994). *A new species of trouble: explorations in disaster, trauma and community.* London: W. W. Norton & Co.

Fortun, K., & Morgan, A., 2016. Thinking across disaster. In Shigemura, J. & Chhem, R. C. (Eds.), *Mental health and social issues following a nuclear accident* (pp. 55–64). Tokyo: Springer.

Fukushima Medical University (2020). Results to date of thyroid screenings (translation ours) (online). Fukushima Health Management Survey. Available at: https://fukushima-mimamori.jp/thyroid-examination/result.html (Accessed: 2 September, 2021)

Fukushima Prefectural Office (2019). Transition of evacuation designated zones (online). Available at: https://www.pref.fukushima.lg.jp/site/portal-english/en03-08.html (Accessed: 2 September 2021)

Fukushima Prefectural Office (2020). Explanation of the changes in evacuation boundaries (translation ours) (online). Available at: https://www.pref.fukushima.lg.jp/site/portal/cat01-more.html (Accessed: 2 September 2021)

Fukushima Prefectural Office (2021). Numbers of evacuees out of the Prefecture as of August 11 2021 (translation ours) (online). Available at: https://www.pref.fukushima.lg.jp/site/portal/ps-kengai-hinansyasu.html (Accessed: 2 September 2021)

Gill, D. A. (2007). Secondary trauma or secondary disaster? Insights from Hurricane Katrina. *Sociological Spectrum, 27*(6), 613–632.

Guggenheim, M. (2014). Introduction: Disasters as politics – politics as disasters. *The Sociological Review, 62*(Suppl. 1), 1–16.

Hasegawa, A., Tanigawa, K., Ohtsuru, A., Yabe, H., Maeda, M., Shigemura, J., Ohira, T., Tominaga, T., Akashi, M., Hirohashi, N., Ishikawa, T., Kamiya, K., Shibuya, K., Yamashita, S., & Chhem, R. K. (2015). Health effects of radiation and other health problems in the aftermath of nuclear accidents, with an emphasis on Fukushima. *The Lancet, 368*(9992):479–488.

Kodama, Y., Oikawa, T., Hayashi, K., Takano, M., Nagano, M., Onoda, K., Yoshida, T., Takada, A., Hanai, T., Shimada, S., Shimada, S., Nishiuchi, Y., Onoda, S., Monma, K., Tsubokura, M., Matsumura, T., Kami, M., & Kanazawa, Y. (2014). Impact of natural disaster combined with nuclear power plant accidents on local medical services: a case study of Minamisoma Municipal General Hospital after the Great East Japan Earthquake. *Disaster Medicine and Public Health Preparedness, 8*(6):471–476.

Leppold, C., Tanimoto, T., & Tsubokura, M. (2016). Public health after a nuclear disaster: beyond radiation risks. *Bulletin of the World Health Organization, 94*(11), 859–860.

Morita, T., Nomura, S., Tsubokura, M., Leppold, C., Gilmour, S., Ochi, S., Ozaki, A., Shimada, Y., Yamamoto, K., Inoue, M., Kato, S., Shibuya, K., & Kami, M. (2017). Excess mortality due to indirect health effects of the 2011 triple disaster in Fukushima, Japan: a retrospective observational study. *Journal of Epidemiology and Community Health, 71*(10), 974–980.

Morita, T., Tsubokura, M., Furutani, T., Nomura, S., Ochi, S., Leppold, C., Takahara, K., Shimada, Y., Fujioka, S., Kami, S., Kato, S., & Oikawa, T. (2016). Impacts of the 2011 Fukushima nuclear accident on emergency medical service times in Soma District, Japan: a retrospective observational study. *BMJ Open, 6,* e013205

National Police Agency (2022). Activities of the police and situation of deaths and injuries after the Tohoku earthquake and tsunami (translation own) (online). Available at: https://www.npa.go.jp/news/other/earthquake2011/pdf/higaijyoukyou.pdf (Accessed: 23 March 2022)

Nomura, S., Blangiardo, M., Tsubokura, M., Nishikawa, Y., Gilmour, S., & Kami, M., Hodgson, S. (2016a). Post-nuclear disaster evacuation and survival amongst elderly people in Fukushima: a comparative analysis between evacuees and non-evacuees. *Preventative Medicine, 82*, 77–82.

Nomura, S., Blangiardo, M., Tsubokura, M., Ozaki, A., Morita, T., & Hodgson, S. (2016b). Postnuclear disaster evacuation and chronic health in adults in Fukushima, Japan: a long-term retrospective analysis. *BMJ Open, 6*(2), e010080.

Norris, F. H., Friedman, M. J., & Watson, P. J. (2002). 60,000 disaster victims speak: part II. Summary and implications of the disaster mental health research. *Psychiatry: Interpersonal and biological processes, 65*(3), 240–260.

Ohba, T., Tanigawa, K., & Liutsko, L. (2021). Evacuation after a nuclear accident: critical reviews of past nuclear accidents and proposal for future planning. *Environment International, 148*, 106379

Perry, R. W. (2007). What is a disaster? In Rodriguez, H., Quarantelli, E. L., & Dynes, R. (Eds.), *Handbook of disaster research: Handbooks of sociology and social research* (pp. 1–15). New York, NY: Springer.

Quarantelli, E. L. (1998). *What is a disaster? A dozen perspectives on the question.* London: Routledge.

Schuller, M. (2016). 'The tremors felt round the world': Haiti's earthquake as global imagined community. In Button, G. V., & Schuller, M. (Eds), *Contextualizing disaster* (pp. 66–88). New York, NY: Berghan.

Shibuya, K., Gilmour, S., & Oshima, A. (2014). Time to reconsider thyroid cancer screening in Fukushima. *The Lancet, 383*(9932), P1883–1884.

Smith, E., Wasiak, J., Sen, A., Archer, F., & Burkle, F. M. (2012). Three decades of disasters: a review of disaster-specific literature from 1977–2009. *Prehospital and disaster medicine, 24*(4), 306–311. doi:10.1017/S1049023X00007020

Sonoda, Y., Ozaki, A., Hori, A., Higuchi, A., Shimada, Y., Yamamoto, K., Morita, T., Sawano, T., Leppold, C., & Tsubokura, M. (2019). Premature death of a schizophrenic patient due to evacuation after a nuclear disaster in Fukushima. *Case Reports in Psychiatry*

The Japan Times (2021). South Korea to start own food service for Olympic athletes on Fukushima fears (online). *The Japan Times.* 17 July, 2021. Available at: https://www.japantimes.co.jp/news/2021/07/17/national/south-korea-olympic-athletes-food/ (Accessed: 2 September, 2021)

Tierney, K. J. (1989). Improving theory and research on hazard mitigation: political economy and organizational perspectives. *International Journal of Mass Emergencies and Disasters, 7*(3), 367–396.

Tierney, K. J. (2007). From the margins to the mainstream? Disaster research at the crossroads. *Annual Review of Sociology, 33*, 503–525.

Tierney, K., & Oliver-Smith, A. (2012). Social dimensions of disaster recovery. *International Journal of Mass Emergencies and Disasters, 30*(2), 123–146.

Tsuda, T., Tokinobu, A., Yamamoto, E., & Suzuki, E. (2016). Thyroid cancer detection by ultrasound among residents ages 18 years and younger in Fukushima, Japan, 2011 to 2014. *Epidemiology, 27*(3), 316–322.

United Nations Office for Disaster Risk Reduction (UNDRR) (n.d.). *Disaster* (online). Available at: https://www.undrr.org/terminology/disaster (Accessed: 27 May, 2021).

United Nations Scientific Committee on the Effects of Atomic Radiation (UNSCEAR) (2010). *Sources and effects of ionizing radiation: UNSCEAR 2008 report to the General Assembly with scientific annexes.* New York, NY: United Nations.

United Nations Scientific Committee on the Effects of Atomic Radiation (UNSCEAR) (2020). *UNSCEAR 2020 report: Sources, effects and risks of ionizing radiation. Annex B: Advance Copy* (online). Available at: https://www.unscear.org/unscear/en/publications/2020b.html (Accessed: 2 September, 2021)

Wind, T. R., & Komproe, I. H. (2012). The mechanisms that associate community social capital with post-disaster mental health: a multilevel model. *Social Science & Medicine*, 75(9), 1715–1720.

Wisner, B., Blaikie, P., Cannon, T., &Davis, I. (2004). *At risk: Natural hazards, people's vulnerability and disasters.* London: Routledge.

Yabuki, S., Ouchi, K., Kikuchi, S., & Konno, S. (2015). Pain, quality of life and activity in aged evacuees living in temporary housing after the Great East Japan earthquake of 11 March 2011: a cross-sectional study in Minamisoma City, Fukushima prefecture. *BMC Musculoskeletal Disorders* 16(1), 1–6.

Part I
Reflections from the Field

2 Reflections from Frontline Healthcare Workers

Katsuka Onoda and Rika Sato

This chapter presents reflections from two nursing professions. The experience of each took a different trajectory at the time of the 3.11 disaster, and the two nurses are colleagues who have spent time discussing their different reactions and experiences. This chapter presents these reflections from the perspective of each writer.

My Ten Years after Experiencing the Great East Japan Earthquake and Experiences from Now On, by Katsuka Onoda

It's already been ten years since that day.

I was born and raised in Minamisōma. I still spend my days living here, hoping to contribute to the community, even just a little, as a nurse.

I was a head nurse that day, and now I'm director of the Nursing Department. And though I experienced the earthquake alongside many members of staff that day, amongst that staff, there are only a few members left with whom I still work today. One valuable staff member who had the same experience as me is Ms. Sato, who is currently the assistant director of the Nursing Department. Even if there are differences in how we felt at the time, in our decisions and choices then, we've each overcome it, and now we've been through thick and thin together, to rebuild the Nursing Department into one that can contribute to the area's medical care.

I'd like to take a look back on my past ten years in the hopes that our experiences can one day prove useful in the future. I wrote a composition as part of a hospital archive created after the Great East Japan Earthquake (one year later). It starts as follows:

That day, I was about to enter my third year since becoming a head nurse.

I had been doing outpatient work as head nurse for three years, and my position immediately after beginning work had also been in outpatient care, so I had been doing it for seven years. That was where I had spent much of my life as a nurse, through working as a single person, getting married, and having children. So I had some attachment to the Outpatient Department. Even if a patient had been coming for treatment

DOI: 10.4324/9781003182665-3

regularly for twenty years, we could look at one another and smile, and find the power to get through anything that was painful or difficult.

But in my days of outpatient work as head nurse, I was depressed because I couldn't do the kind of outpatient nursing that I wanted to do. I worried about how I would face my third year as head nurse, and honestly, I was eager to escape. That was when the Great East Japan Earthquake and nuclear disaster occurred.

Then at the end, to describe my feelings one year afterward, I concluded my writing as follows:

In the end, I love nursing, and Minamisōma Municipal General Hospital is dear to me.

And maybe I can be a little more valuable by taking on the role of nurse. A head nurse cannot fulfil their role without the staff. If the chief, the staff, and the patients didn't need a head nurse, there would be no such role. I don't know what exactly it is I need to do. I think there were probably staff members who had been hurt by things I'd said as head nurse at the time of the earthquake. I'm full of regret. But that day won't come again, and I don't want it to. So, I want to look forward and think of what to do, from now on, as a nurse, as head nurse. ...

At the time of the earthquake, I'd had 18 years of experience as a nurse. As a nurse, as a mother, as a wife ... I didn't do any one of these roles enough justice, but I lived them in my own way, holding on for dear life. I wanted to run away just from the difficulty of doing so. But from that day forward, I was reminded of just how irreplaceable and important each of these roles was to who I was.

When I was a nursing student, I started thinking about what kind of nursing I wanted to do. I thought it was necessary to learn as a public health nurse in order to do so, and so I went on to higher education. I had always thought that I wanted to contribute to the area in which I was born and raised.

So, perhaps the reason I chose to stay at the hospital that day wasn't because I had some strong conviction but just because it was a continuation of my daily life and I couldn't think much more deeply than that. So, seeing the state of my children after the earthquake, hearing their stories, as a mother I felt deep regret that I made them feel lonely, including the choices I made that day. I had been able to continue my job as a nurse precisely because I had been told, 'Mum, do your best'. I was able to reaffirm that my family was more important than anything to me.

I've had ten years of experience as a nurse since the earthquake. Three years after becoming Assistant Director, I realized the reality of this area at Odaka Hospital, which was unable to reopen its doors after the earthquake. I think I was able to make a small contribution to community-based health care focused on home medical care, all the while thinking about what I could do now as well as for the future. This experience also allowed me to reaffirm how I want to live – not just as a nurse, but for myself. These two years at Odaka have formed the foundation for who I am ten years after the earthquake and who I will be from the 11th year onward.

To be honest, ever since I became Assistant Director, I had been in self-denial every day. It was especially painful to look back on my experience of the earthquake, because from the earthquake until now, I had thought that talking about what I experienced at the hospital after the nuclear accident might hurt someone else. I hadn't had the opportunity to share my experience with others who'd had different experiences, and I now think that perhaps I needed time to be able to talk calmly about what I had gone through, to think of it as an 'experience'.

After I came back to the general hospital as Director of the Nursing Department in 2019, I was finally able to share my experience. And now, along with many members of staff who have not had the same experience, I manage nursing even as I struggle day to day.

Even as I rose from nursing student to Director of the Nursing Department, my view of nursing, that it should stay close to the patient, has never changed. Even so, I've been asked by nurses who came to lend their support, 'What do you mean by 'stay close'?' I wasn't able to give a good answer from just my experience of the earthquake. But from my experience at Odaka, I had come to believe that 'staying close' meant using nursing to support others' ability to be themselves.

I don't know if this idea is correct. But for me, who believes that there is nursing I can practice precisely because I experienced the earthquake, from now on, I want to live my own life in the roles I've been given without comparing myself to anyone else. And I want to take on the difficult task of cultivating human resources while cherishing the bonds I have with those who have worked hard with someone like me. If that could become a foundation for the future one day, I couldn't be happier.

I wonder what a future me, ten years from now, would think if she read these words. When I think about it, I feel sincerely grateful I was given the opportunity to do so.

The Ten Years Since Then, and the Next Ten Years, by Rika Sato

That day, I was working a day shift in the ward, having finished assisting a bedridden patient with bathing, helping them get dressed. Because the ward was on the sixth floor, the shaking was quite strong, and I still vividly remember thinking that the hospital would collapse, just like that. At the time, I was the ward's head nurse.

It's been ten years since then, and when I look back on that time, it pains my heart.

After the huge earthquake and tsunami on March 11, large aftershocks were frequent. Among the staff, there were those whose homes had been washed away by the tsunami and others whose homes and families were swept away. It wasn't just hospitalized children and families who were crying; some

of the staff also burst into tears. The inside of the ward was flooded by rup-
tures in the water pipes, and you could see the approaching tsunami through
the windows. It was just chaotic. Meanwhile, several staff members, including myself, rather than return home, worked continuously day and night for
several days, taking turns to rest. Then, on March 12, the hydrogen explosion
at the Fukushima Daiichi Nuclear Power Plant Unit 1 building completely
transformed our existing anxiety into anxiety mixed with the fear of death.
The Fukushima Daiichi Nuclear Power Plant incident continued after that
with Units 2, 3, and 4, increasing our fear day after day.

On March 15, after the night shift had ended, I left the hospital and evacuated to Yonezawa in Yamagata Prefecture, with the thought that I just had to
evacuate my children (who at the time were in fifth, fourth, and first grade of
elementary school). Naturally, I had planned to return to the hospital right
away. It was a natural decision for me, thinking of the staff and the head of
the ward, who had been working hard with me up to then. However, when I
saw my children for the first time since the earthquake, they looked relieved
and said sadly to me, 'Don't go'.

At first, I would have had no hesitation to return to the hospital, but this
feeling immediately changed into a heavy doubt. On top of that, while I was
prepared to die if I were in Minamisōma, my husband was also unable to
leave Minamisōma due to work. My brother's words burned into my ears:
'Right now, you're all the children have. What are they going to do if both of
their parents die?' I gave up going back to the hospital. It was undoubtedly a
bitter choice for me to make as a member of society, as a nurse, and as a
mother.

That choice soon turned to guilt and regret. My own irresponsible actions,
the fear of the staff who remained there, and the work of evacuating patients –
the thoughts were so strong that they were almost tangible and truly tore me
apart. Meanwhile, my children were near me, full of relief. Those days were
unbearably complicated.

After that, I worked at evacuation centres in Yonezawa, Yamagata Prefecture,
and Nagaoka, Niigata Prefecture. From my experiences during that time, I
learned and thought about a lot, and it had a great influence on my current
style of nursing and my view of nursing in general. At each evacuation centre
in and outside the prefecture, I received a wealth of support and warm wel-
comes from the evacuees. I was keenly aware of it both as an evacuee myself
and as a nurse working in the shelter, and even now I'm filled with gratitude.
Under these circumstances, citizens of Minamisōma said of having local
nurses nearby: 'I feel relieved'', 'I can speak my mind', 'Even with the same
nurse, something's different', 'Thank you for coming with me'. They taught
us that our existence had value. And if there was any problem, we would
think about it together with the evacuees and try to find clues to a solution to
deal with it. I can never forget how it felt at the time to realize how I was supported by the citizens of my hometown of Minamisōma.

Then, starting in June 2011, I returned to work at Minamisōma Municipal
General Hospital. At the time, there was a period in which I was seriously

thinking about getting a job at an evacuation centre, but when I returned to work, I felt happy and proud to be able to work as a nurse at Minamisōma Municipal General Hospital. Plus, I was grateful for and respected everyone who remained at the hospital at the time and had responsibly performed their duties. Those feelings have never changed.

All my experiences from the Great East Japan Earthquake have significantly changed my outlook on life, family, and nursing. At that time, I vowed to myself that I would always take responsibility for my work in any situation. I also told the children that, and I'm not sure if they really understood, but they accepted it. My mother also promised to help. Of course, even now, nearly ten years later, those things I vowed have not changed.

'There's no such thing as a useless experience, because every experience you have has made you who you are.' These words touched my heart at the time and supported me. Most of my experiences are painful to remember, but all of them are important to deepening my outlooks on life, family, and nursing, and they should be great sources of encouragement for my life.

Then there's the next ten years. At the time I was 40 years old, and soon I'll be 50. That's about ten years from retirement age. Of the ten years of experience I've had, a particularly large part of it was me being supported by the locals, and I want to give back to them. Although I have little power, I have a strong desire to contribute to the community, and I think that what I can do is to be involved in medical care and nursing and to keep aiming toward nursing that can provide peace of mind, safety, and satisfaction through the entire hospital and the entire region. I want to do my very best for the rest of my life as a nurse so that the local residents can live with peace of mind.

Minamisōma was formed by merging one city and two towns in 2006 and is composed of three wards: Haramachi Ward, Kashima Ward, and Odaka Ward. Of these, Odaka Ward is located within 20 kilometres of the Fukushima Daiichi Nuclear Power Plant, and the nuclear accident caused by the Great East Japan Earthquake forced all its residents to evacuate. They weren't allowed to return home until the evacuation order from the government was lifted, a period of five years four months. While what was then Minamisōma Municipal Odaka Hospital cannot perform the same functions it did before, it now supports regional care in Odaka Ward as Odaka Clinic, in association with the Municipal General Hospital. For the residents who had experienced so many hardships and returned to Odaka Ward, the existence of the Odaka Clinic must be a great relief. I think I want to continue to improve medical care in the area in order to continue to provide that peace of mind.

Due to the phenomena associated with an ageing society, the need for comprehensive community-based systems, such as home nursing, are also increasing, and the role of nurses who are directly involved in providing guidance for the lives and health of local residents is incredibly important, as is represented by in-home nursing and medical care. Even for our hospital, I think it's necessary to not be limited to nursing at the hospital but to actively go out and lead the charge towards improving home nursing. I want to value the voices of those who are being treated and living at home. Additionally, I want to further maintain and improve the health of residents in the entire region by creating an environment that enables closer cooperation with other hospitals and clinics in the area.

I want to make an effort to make the next ten years such that when I look back on them ten years in the future, I can think to myself, 'I was able to contribute to the local people. I was able to give back, even a little bit.'

3 Psychiatric Care after the Nuclear Disaster in Fukushima

Arinobu Hori

I am a psychiatrist from Tokyo who voluntarily moved to Minamisōma City, Fukushima Prefecture, in 2012, following the disaster. I recognised the need for inpatient care and worked for three years to reopen the sole psychiatric hospital in Minamisōma City, which was temporarily closed due to the disaster. I then opened my own mental health clinic in the city in 2016. While responding to the needs for general psychiatric care in the disaster-affected areas, I noticed that some of the patients with prolonged depression and anxiety disorders had recurrent flashbacks and nightmares about the disaster, expressing a state of high terror. Therefore, I established a context-specific system to provide cognitive-behavioural therapy for post-traumatic stress disorder (PTSD; Hori et al., 2016; Hori et al., 2018; Hori et al., 2019; Hori, Ozaki et al., 2020a; Hori, Takebayashi et al., 2020b; Hori et al., 2021). I also provide PTSD treatment in the disaster-affected areas with the support of the non-profit organisation People's Hope Japan. This chapter notes my thoughts based on nine years of experience working in the disaster area as a clinician.

Disasters reveal structural weaknesses that exist in society and disproportionately affect vulnerable populations. People with severe mental health disorders can be one of the most vulnerable groups in society. The Great East Japan Earthquake, tsunami and nuclear disaster may have had a considerable impact on people with mental health disorders, but there has not been enough systematic investigation or research on this topic. This chapter provides an overview of psychiatric issues in Fukushima Prefecture after the disaster, including reflections not yet accompanied by a scientific evidence base. The need for further evidence is an important issue in this context. The experiences and health consequences of the disaster on people with pre-existing mental health disorders is an underexamined issue.

From the perspective of a post-disaster mental health service provider, I think it is advisable to focus on two groups of people as the target populations to be served. There are different questions we can ask in relation to each of these groups:

(1) How to continue to meet the needs of people with diagnosed mental health disorders who require special care even during non-crisis times

DOI: 10.4324/9781003182665-4

(mainly those with schizophrenia, bipolar disorder and severe intellectual disabilities) without significant interruption during disasters. In addition to inpatient and outpatient psychiatric treatment, medical support such as daytime care and home nursing can be important to support the community life of people living with mental health disorders, and welfare facilities such as employment support facilities are also important.

(2) How to identify and respond to the needs of people who did not previously experience mental health symptoms pre-disaster but need therapeutic intervention for symptoms brought on by the prolonged and intense stress of the disaster.

With respect to the first group, there were five psychiatric hospitals providing medical care for people with mental health disorders in the present disaster area before the earthquake, mainly providing inpatient facilities (Kumakura, 2011). These five hospitals were located within a 30-km radius of the Fukushima Daiichi Nuclear Power Plant and were therefore within the initial evacuation zone. This necessitated the evacuation of inpatients after the disaster, with a total of 713 patients evacuated (breakdown across hospitals: 180, 104, 56, 339, and 34 patients, respectively; ibid.). Subsequently, Gotoh et al. (2021) reported that the Standardised Mortality Ratio of all psychiatric inpatients who were evacuated immediately from hospitals located within the government-designated evacuation zone in 2011 was as high as 3.33 (for all psychiatric inpatients) and 5.43 (for inpatients evacuated within Fukushima Prefecture). In other words, mortality rates were over three times higher and five times higher for these groups, respectively, than generally expected mortality rates. However, exactly what happened during the evacuation of psychiatric inpatients immediately after the nuclear power plant disaster remains unknown in many respects. It is important to continue gathering knowledge about individual cases, in addition to overarching data, in order to better grasp the whole picture. To fill this gap, it is necessary to conduct case studies and integrated research in the future in which epidemiological data are interpreted in conjunction with findings from the social sciences and humanities.

From my experience of working as a psychiatric clinician since 2012 and managing an inpatient ward, there was only a short period (a few months) where we saw an increased need to accept patients with worsening schizophrenia and other mental illnesses, after reopening the inpatient ward. There are two hypotheses that might explain this. One hypothesis is that people with mental health disorders were struggling to cope with life in the disaster-affected area and had already moved to neighbouring areas by the time the ward was reopened in April 2012. The other hypothesis is that teams from Fukushima Prefecture and Fukushima Medical University had actively engaged in providing psychiatric services, mainly through outreach programmes. Volunteer activities from all over Japan greatly contributed too. These teams had a firm intention to provide mental health care in the disaster-affected areas, where five psychiatric hospitals were no longer available and the situation was critical. Such activities have been substantially effective but are unlikely to remain as long-term resources. Still, these two hypotheses have yet to be fully examined and the reason for the unexpected trend in inpatient admissions remains unknown.

Even people with no history of mental health disorders before the disaster faced significant stress from the earthquake, tsunami, and nuclear disaster. However, within this group, certain populations are more likely to manifest explicit mental health symptoms. For example, one such population is older adults (Hori, Ozaki et al., 2020a). In the psychiatric ward where I worked, most inpatients were older adults whose dementia had worsened after the disaster and who developed severe behavioural abnormalities. The changes in their living environment, such as being forced to evacuate, being unable to continue daily routines of farming or interacting with close acquaintances and friends, and changes in their family structure due to their children and grandchildren evacuating to distant places, were beyond what many older people could handle. In this situation, behavioural and psychiatric symptoms of dementia such as wandering and agitation appeared; many patients needed support and ended up being admitted to psychiatric wards. Japanese facilities for older adults are not equipped to deal with severe behavioural abnormalities, and it is not uncommon for these patients to be admitted to a psychiatric hospital.

The impact of the nuclear power plant disaster on the mental health of mothers and children has also been quite significant. Goto et al. (2015) found that 27.6% of mothers who were pregnant at the time of the Fukushima nuclear disaster exhibited depressive symptoms. I am not a child psychiatrist and have not been involved in this topic much, so I will not go into further detail in this chapter, but I believe this is one of the most critical issues.

Even among the general adult population who might have a high tolerance for stress, many people were overworked while contributing to post-disaster reconstruction and then experienced depressive symptoms (Hori et al., 2016; Hori, Takebayashi et al., 2020b). There were also people who had PTSD symptoms related to the disaster that persisted but went unrecognised only to become recognised cases years later (Hori et al., 2018; Hori et al., 2019; Hori, Takebayashi et al., 2020b; Hori et al., 2021). This indicates the need for early evaluation and monitoring of mental wellbeing in the population following a disaster.

The Great East Japan Earthquake and the nuclear power plant disaster have enormously affected the local population. Based on documented experience after the Chernobyl disaster (Bromet, 2012), we can expect that the Fukushima nuclear disaster would have had a severe effect on the local population. To unravel the phenomenon of mental health after the Fukushima disaster, it is necessary to integrate narrative approaches and empirical studies. In addition to the usual epidemiology and psychiatry studies, I believe an integrated approach that includes the entire range of social sciences and humanities will be indispensable. However, due to the lack of researchers and clinicians, many issues have not been sufficiently investigated. It is necessary to continue to address these issues and to convey the importance of this field of practice to society as a whole and call for further support.

References

Bromet, E.J. (2012) Mental health consequences of the Chernobyl disaster. *Journal of Radiological Protection*, Vol. 32 (1), pp. N71–75. doi:10.1088/0952-4746/32/1/N71.

Goto, A., Bromet, E.J. and Fujimori, K. (2015) Immediate effects of the Fukushima nuclear power plant disaster on depressive symptoms among mothers with infants: a prefectural-wide cross-sectional study from the Fukushima Health Management Survey. *BMC Psychiatry*, Vol. 15 (1), pp. 59–59. doi:10.1186/s12888-015-0443-8.

Gotoh, D., Kunii, Y., Terui, T., Hoshino, H., Kakamu, T., Hidaka, T., Fukushima, T. and Yabe, H. (2021) Markedly higher mortality among psychiatric inpatients mandatorily evacuated after the Fukushima Daiichi Nuclear Power Plant accident. *Psychiatry and Clinical Neurosciences*, Vol. 75 (1), pp. 29–30. doi:10.1111/pcn.13158.

Hori, A., Hoshino, H., Miura, I., Hisamura, M., Wada, A., Itagaki, S., Kunii, Y., Matsumoto, J., Mashiko, H., Katz, C.L., Yabe, H. and Niwa, S.-I. (2016) Psychiatric outpatients after the 3.11 complex disaster in Fukushima, Japan. *Annals of Global Health*, Vol. 82 (5), pp. 798–805. doi:10.1016/j.aogh.2016.09.010.

Hori, A., Morita, T., Yoshida, I. and Tsubokura, M. (2018) Enhancement of PTSD treatment through social support in Idobata-Nagaya community housing after Fukushima's triple disaster. *BMJ Case Reports*, Vol. 2018 (2018-06-19), p.bcr-2018-224935. doi:10.1136/bcr-2018-224935.

Hori, A., Ozaki, A., Murakami, M. and Tsubokura, M. (2020b) Development of behavior abnormalities in a patient prevented from returning home after evacuation following the Fukushima nuclear disaster: case report. *Disaster Medicine and Public Health Preparedness* 2020-07-04, pp. 1–4. doi:10.1017/dmp.2020.158.

Hori, A., Sawano, T., Ozaki, A. and Tsubokura, M. (2021) Exacerbation of subthreshold PTSD symptoms in a Great East Japan Earthquake survivor in the context of the Covid-19 pandemic. *Case Reports in Psychiatry*, Vol. 2021 (2021-02-10), pp. 1–3. doi:10.1155/2021/6699775.

Hori, A., Takebayashi, Y., Tsubokura, M. and Kim, Y. (2019) Efficacy of prolonged exposure therapy for a patient with late-onset PTSD affected by evacuation due to the Fukushima nuclear power plant accident. *BMJ Case Reports*, Vol. 12 (12), p. e231960. doi:10.1136/bcr-2019-231960.

Hori, A., Takebayashi, Y., Tsubokura, M. and Kim, Y. (2020a) PTSD and bipolar II disorder in Fukushima disaster relief workers after the 2011 nuclear accident. *BMJ Case Reports*, Vol. 13 (9), p. e236725. doi:10.1136/bcr-2020-236725.

Kumakura, T. (2011) Evacuation of psychiatric inpatients at the time of the nuclear accident in Fukushima Prefecture. *Rinsho Seishin Igaku*, Vol. 40 (11), pp. 1417–1421. (In Japanese).

4 Fukushima Hamadōri (Coastal Area) High School Academy

Learning and Understanding about Nuclear Disaster with Fukushima High School Students

Yasuhiro Takamura

Introduction

I am a high school teacher at Fukushima Hamadōri High School Academy. Since the Fukushima disaster, I have been involved in activities to support students. Despite progress in decontamination and decommissioning work following the Fukushima disaster, local residents, particularly young people in Fukushima Prefecture, have experienced prejudice and stigma. This chapter describes our activities to support local students to overcome such prejudice and recover their confidence through excursions to other sites affected by nuclear disasters, including areas of Belarus affected by the Chernobyl disaster and areas of the U.K. affected by the Windscale fire.

After the Fukushima disaster in 2011, there was unreasonable discrimination against people associated with the Fukushima region. This included refusal of entry into taxis or vandalism of Fukushima license plates due to public speculation in other prefectures that Fukushima residents were radioactively contaminated and might disperse radiation around them. There was also bullying of evacuees from Fukushima outside the prefecture and rumours that unmarried women from Fukushima Prefecture would not be able to marry men in other prefectures.

It is expected that the ongoing reconstruction and decommissioning work caused by the disaster at Fukushima Daiichi nuclear power station, run by Tokyo Electric Power Company (TEPCO), will be passed down to younger generations. Happy Road Net, a non-profit corporation, has provided opportunities and activities for high school students in Hamadōri, Fukushima Prefecture, since before the disaster. Since the disaster, they have provided radiation education and nuclear energy cycle education for high school students and have contributed to town development projects. I planned and ran one such development project to help high school students acquire accurate knowledge regarding radiation and protect themselves from exposure to radiation and stigmatisation.

After the 3.11 disaster, we connected with other communities that had been affected by nuclear disasters. The Republic of Belarus, which experienced fallout from the Chernobyl Nuclear Power Plant disaster in 1986, has been accepting junior and high school students from Fukushima every summer vacation

DOI: 10.4324/9781003182665-5

since 2012 at Zubryonok, the Minsk region's national education and recuperation centre. Whilst in Belarus, we also decided to visit the reconstruction area of Khoyniki, Gomel Region, which had been contaminated by radiation from the Chernobyl nuclear disaster. In addition, Sellafield Ltd. (formerly Windscale and Calder Works) in the United Kingdom, which reprocesses spent nuclear fuel from nuclear power plants, is cooperating with TEPCO. We also decided to visit the area around Sellafield with TEPCO's support.

Each year, the number of participating students from Hamadōri was around 20 from the second year of high school. Operating expenses such as travel and accommodation were covered by reconstruction-related grants from related ministries and organisations, as well as donations from local companies. Two advance learning sessions were held before travel, and post-learning and debriefing sessions were held after returning to Japan.

Overseas Study in Belarus

We visited Belarus twice, in the summer of 2017 and 2018. Since we adjusted our plans for 2018 based on our reflections from the visit in 2017, I briefly explain the details of the visit in 2018.

In advance learning sessions before travelling, the students were informed of the outline of the Fukushima Daiichi nuclear power station disaster, the subsequent reconstruction situation, and the details of the Belarus visit in 2017. In addition, we divided into groups to prepare presentations on the current situation in Fukushima and Japanese culture for local students and residents in Belarus.

We visited Belarus from 23 July to 3 August 2018. From the 25th to the 26th, we visited Praleska, a children's recreation facility in Gomel Region. This centre accepts children and students from areas with high radioactive contamination due to the Chernobyl disaster and has a well-developed health programme. We stayed there for one night and two days to interact with local children and students. In addition to the facility tour, we introduced each other's cultures, played games from Praleska together in groups and danced. What impressed us especially was the comprehensive health care for children, covering not only internal radiation exposure measurements and thyroid screenings but also dental treatments, programmes for obesity and stress management.

On the morning of 27 July, we heard from the director and deputy director of the Gomel State Executive Committee in Gomel City about the response to the Chernobyl disaster. In the afternoon, we heard from an expert who responded to the Chernobyl accident at Gomel University. As a result of the Chernobyl disaster, a policy of mass migration from high-radiation areas was put into place, and villages in high-radiation areas were abandoned. The contaminated area has now been decontaminated as much as possible, and new town development is being carried out, with companies being invited back into the area. It is said that there are generous financial guarantees (higher salaries and better social welfare coverage) for those who work in the previously contaminated areas.

On the morning of 28 July, we visited the Khoyniki District Museum to see the tools used and work done following the Chernobyl disaster, as well as photos of the abandoned villages. We then visited the gate of the high-dose Polesie State Radioecological Reserve. In the afternoon, our students interacted at Strelichevo Secondary School, which was established near the gate to the reserve. At this school, we learned about radiation education and radiation food inspections that are conducted at the school, and the Japanese students introduced Japanese culture to local students. The students at Strelichevo have received radiation education since primary school, and they showed a solid understanding of radiation. Such education may be unique to the disaster area, but it is one of the education systems that we would like to use as a model in Fukushima Prefecture.

The next day, we visited the Khatyn Memorial Complex for the Khatyn massacre during World War II, and in the afternoon, we visited the Belarusian State Museum of the History of the Great Patriotic War. We learned not only from the perspective of responding to the nuclear disaster but also from the perspective of how to impart the tragic events of war to future generations. On the 30th, we visited Nesvizh Castle, a World Heritage site, and learned about its history. In the morning of the 31st, we introduced Japan to the local citizens at a hotel in Minsk, and in the afternoon, we held a Japanese cultural experience (origami, calligraphy, yukata dressing) and a Soran Festival at a shopping centre. We felt that we were playing a part in the cultural exchange between Japan and Belarus. On 1 August, we heard from the ambassador of the Embassy of Japan in Belarus about the relationship between Japan and Belarus, and then we divided into groups and toured the city under the guidance of students majoring in Japanese at Belarus State University.

After returning to Japan, we conducted post-learning and held a debriefing session by inviting relevant ministries and agencies from which we had received grants, as well as various organisations and local companies that had donated to our trip.

Overseas Study in the United Kingdom

From 6 to 16 August 2019, we took a new group of students to the United Kingdom. In the advance learning session, students learned about the outline of the Fukushima Daiichi nuclear disaster, the subsequent reconstruction situation and the 1957 Windscale fire accident at Sellafield Nuclear Power Plant (formerly Windscale and Caldar Works) in the United Kingdom. In addition, we divided into groups to practise presentations that reported to local British students and residents on the current state of reconstruction in Fukushima.

On 8 August, we visited Sellafield Ltd. in Cumbria. In the morning, we entered the Calder Hall reactor, which is undergoing decommissioning, and looked at the reactor control room and various inspection rooms. We also entered the Thermal Oxidation Reprocessing Plant (THORP) and saw the spent nuclear fuel pool and the cask being carried in for reprocessing. In the afternoon, we visited the training centre and observed their training operations.

The next day, we visited a local school, the West Lakes Academy (WLA), in collaboration with Sellafield Ltd. We received a WLA school tie and, in exchange, we gave a presentation about Japanese culture and the current situation in Fukushima. We visited the school building and said farewells. In the afternoon, we were given a guided tour of the National College for Nuclear facility (NCfN). Both the WLA and the NCfN have well-established learning environments and we felt their enthusiasm for education.

On 10 August, we visited the Beacon Museum, using its meeting space to look back on our studies of the past two days. We also visited the second floor of the museum, where the 'Sellafield story' is displayed. On the 11th, we travelled from Whitehaven, Cumbria, to London. On the morning of the 12th, we had a sightseeing trip to Cambridge. In the afternoon, we visited Dojima Sake Brewery, which makes sake at Fordham Abbey on the outskirts of Cambridge. We toured the sake brewery and received a lecture from Vice President Kiyomi Hashimoto about Japanese people and business overseas. It was a valuable experience to see overseas sake breweries and manor houses and listen to the stories of leading people who are active overseas.

On the morning of 13 August, we visited the Embassy of Japan in London and heard from the then ambassador, Koji Tsuruoka, about the relationship between Japan and the United Kingdom. After that, we had lunch at the cafeteria of University College London (UCL) and received a lecture from Professor Shinichi Onuma of the UCL Institute of Ophthalmology, who is from Fukushima Prefecture. It was a valuable experience for our students to use the cafeteria and take lectures at an overseas university. In the evening, at St. James Church in Central London, we introduced the reconstruction situation in Fukushima to local people mainly from the Fukushima Prefectural Association in London. We also hosted a Japanese cultural experience (origami, calligraphy, yukata dressing) and a Soran Festival in London, as we had done in Belarus. On the 14th, we went sightseeing in groups under the guidance of the Fukushima Prefectural Association in London.

After returning to Japan, post-learning was conducted, and a debriefing session was held, as we had done following the Belarus study.

Domestic Study in Rakkasho Village

The visit to Rokkasho Village was not originally planned, but after studying the nuclear fuel cycle at Sellafield Ltd., the students suggested that they would like to know about the nuclear fuel cycle in Japan. Therefore, on 22 September 2019, we decided to visit Japan Nuclear Fuel Limited (JNFL) in Rokkasho Village, Aomori Prefecture. First, at the Rokkasho Visitor Center, we heard about the nuclear fuel cycle and the reprocessing process. After that, we visited the observation room of the Low-Level Radioactive Waste Disposal Centre and then toured the inside of the reprocessing plant. Here, we visited the central control room, spent nuclear fuel pool and high-level management facility, and we took a bus tour around other facilities of the reprocessing plant. The site of JNFL seemed well-organised, just like the reprocessing facilities at Sellafield Ltd.

Conclusion

At Sellafield, we explored universally relevant issues connected to nuclear power, energy and the conflict among science, technology and society. The Calder Hall nuclear power station is one of the oldest commercial nuclear power plants in the world and is currently in the process of decommissioning after being in operation from 1956 to 2003, including the period of the Windscale fire disaster in 1957. Sellafield's THORP has received and reprocessed a large amount of spent nuclear fuel from Japanese nuclear power plants. At Sellafield, we learned about the history of this relationship between nuclear power plants in Japan and the United Kingdom up to the present day. We also learned about previous cases of nuclear disasters and tensions in society about nuclear waste disposal.

High school students from Hamadōri, Fukushima Prefecture, visited Sellafield and the surrounding area to learn about the nuclear disaster, decommissioning, nuclear waste treatment and conflicts with society. The students then deepened their learning at the Reprocessing Plant in Rokkasho Village, Aomori Prefecture. In doing so, they honed their thinking. The United Kingdom had different lessons than Belarus. Sellafield has had a history of coexistence with the 'nuclear disaster' and 'decommissioning' for more than half a century, and there is a process to complete decommissioning work which looks toward the next 100 years or more. Education and learning were enriched so that people and communities could grow, and there was a model of how industry and government were involved.

Ten years have passed since the Great East Japan Earthquake and the disaster at TEPCO's Fukushima Daiichi nuclear power station. In some parts of the disaster-affected area, new buildings have been built and reconstruction is underway, while in other areas, it is still not possible even to return home. The road to reconstruction is long and steep, and the change of generations continues. We must bear the future of our hometown, even though we know it will be a burden to the younger generation. I hope that the participating high school students from Hamadōri will protect themselves from both radiation and stigmatisation and contribute to future town development, using the knowledge they have gained in Belarus, the United Kingdom and Rokkasho Village.

5 The Increased Disaster-Related Deaths after the Fukushima Nuclear Disaster and the System for Their Compensation

Shigeta Mimura

Introduction

From my standpoint as a lawyer who has been working in Fukushima, in this chapter I consider the health and social issues that occurred in the prefecture, including disaster-related deaths and the legal problems associated with the disaster.

The number of disaster-related deaths from the Great East Japan Earthquake (those who did not die directly from the earthquake or tsunami but died of illness due to the physical burden of life after evacuation, etc., and were determined to have died because of the disaster) reached a toll of 3,768, of which 2,314 constituted disaster-related deaths in Fukushima Prefecture (Reconstruction Agency, 2021). This number is significantly larger than that of Miyagi and Iwate Prefectures, which were heavily damaged by the tsunami. This reflects the fact that many residents of Fukushima who had evacuated due to the nuclear accident fell ill along the way and died. In addition, from my own gut feeling and professional experiences, having come into contact with victims, the actual impact not included in the published figures cannot be underestimated. There were quite a few cases involving those who had lost someone who had become ill and died shortly after evacuation but who were hesitant to register their loved ones as disaster-related deaths. There were also cases in which people died while evacuating from "voluntary evacuation zones" such as Iwaki and Koriyama; these were not recognized as disaster-related deaths because they had occurred outside the areas ordered by the government to be evacuated. Then, even if they did not lead to death, there were many cases in which the conditions of patients with chronic illnesses (whose symptoms had been stable) worsened immediately after living in evacuation and in which patients' symptoms worsened due to the inability to perform adequate rehabilitation.

Many of the legal disputes I have been involved with in Fukushima arose with the nuclear disaster as a trigger, for example the intermediate exploitation in decontamination work in areas contaminated by radioactive materials (the number of decontamination workers at peak hours was about 35,000) and the fear of increased crime due to deteriorating security; labor-related incidents during decommissioning work (the number of workers at the

DOI: 10.4324/9781003182665-6

Fukushima Daiichi Nuclear Power Plant during peak hours in March 2015 was 7,450 (Tokyo Electric Power Company Holdings, Incorporated, 2017); trouble in large-scale development activities meant to solve shortages in temporary housing for evacuated residents and housing at evacuation destinations; construction-related disputes regarding houses purchased at evacuation sites; claims for compensation from farmers and travel agents suffering reputational damage from rumors surrounding radioactive materials; the bankruptcy of a factory for precision machinery whose supply chain had been disrupted by the failure of parts manufactured in Fukushima to be delivered; a divorce case involving a separated family (the father stayed behind at the company he worked for at the time of the nuclear accident while only the mother and children left Fukushima and evacuated far away for fear of radioactive harm); inheritance division cases surrounding compensation money; cases of fraud in which donations were solicited by those posing as a public organization; trouble with landowners regarding reconstruction work of areas with evacuation orders; suicide caused by bullying of children who had evacuated from Fukushima; and so on.

In the legal field, there is a tendency to pay attention to litigation involving residents seeking compensation for nuclear damages from the national government and Tokyo Electric Power Company Holdings, Incorporated (TEPCO), or litigation which sought to prohibit nuclear power plants in various parts of Japan (which had shut down after the Fukushima Daiichi accident) from restarting. But there were also many other legal issues which came about in Fukushima Prefecture following the disaster. However, the number of lawyers working in Fukushima Prefecture (with a population of about 2 million), where the nuclear incident occurred, is only about 200, and there has been a lack of human resources to deal with proceedings for compensation for damages based on each individual's circumstances and to meet their legal needs. In order to rebuild the lives of evacuated residents, as well as to rebuild the region, establishing a system of consultation and means of dispute resolution in order to properly resolve these issues in accordance with the law is a major challenge.

Study Sessions with Doctors

I responded to residents' requests for legal advice not only in the meeting room of a law firm in Iwaki but also at meeting places in temporary housing at evacuation sites, at the Japan Legal Support Center (Houterasu), counseling rooms at city halls, and so forth. I was often confronted about the harsh realities of evacuation due to the nuclear accident, and the dire situation of disaster-related deaths arising from poor physical conditions. I heard stories such as that of a father who had lost consciousness at the gymnasium to which he had been evacuated, was taken to the hospital by ambulance, but died a week later; there was also a wife who had become depressed and alcoholic during life in evacuation and committed suicide shortly after her family had temporarily returned home near the Fukushima Daiichi Nuclear Power Plant.

In order for these bereaved families to receive appropriate compensation for disaster-related deaths, it is necessary to prove that there is a causal relationship between evacuation from the nuclear accident and the cause of death. This involves asking for medical records of deceased patients from the hospitals in areas that had been evacuated or at evacuation destinations, where the patients had either been hospitalized or had visited, as well as for documents to be submitted to government offices. This requires checking a huge amount of medical documentation, and there are many difficulties involved. Of course, since TEPCO spends a large amount of money on legal fees to collect materials and makes claims that deny such causal relationships (Ministry of Economy, Trade and Industry, 2012), the bereaved families must prove that they can withstand this.

Other lawyers and I were fortunate to meet with doctors who had been active in Fukushima Prefecture since the nuclear accident. Beginning in 2015, we were given the opportunity once every three to four months to hold study sessions to break down and analyze the contents of the medical records of patients who had died because of the disaster. These study sessions were very meaningful for Fukushima lawyers, who had been struggling with the issue of disaster-related deaths. Receiving proper information and advice concerning medical records from the experts (doctors) was incredibly useful to the lawyers' job of establishing proof, and there were many cases in which compensation was actually granted and the bereaved families were helped. I would like to take this opportunity to thank those doctors for all their cooperation.

The most common cases discussed during the study sessions were deaths from cardiovascular disease during the evacuation, as well as from pneumonia due to weakened immune functions and poor oral management. It seems that many evacuees were unable to take medicines for conditions such as high blood pressure and diabetes because they had evacuated without taking enough of their belongings; they believed that they would be able to return home in a few days. In addition, they were forced to sleep on the cold floor of a school gymnasium or assembly hall at their evacuation destination without getting enough to eat, and so older adults with weakened physical strength and immune functions incurred damage to their health.

Even for those who moved into apartments at their evacuation destination, there were many cases in which people lost their local communities, which had supported the foundation of their former lives, and their physical conditions deteriorated due to the inability to respond to such sudden changes in their living environment. According to a survey conducted by NHK in 2019 (Japan Broadcasting Corporation, 2019), the average number of evacuations per evacuee was 6.7 times, which shows that they were moving around frequently. For instance, there was a family in Okuma, where the Fukushima Daiichi Nuclear Power Plant is located, who evacuated to a public hall in a neighboring town. From there, they took a bus to a shelter in the neighboring town, but the shelter was full of people and they were unable to secure enough space to sleep, so they evacuated to the gymnasium of a nearby school. They then traveled by bus to Big Palette Fukushima (a convention center to which about 2,500 people evacuated), relatives' home in Tokyo, temporary housing in the Aizu region, an apartment in Iwaki, and public housing for recovery built in Fukushima Prefecture, repeatedly moving around due to evacuation. As the nuclear accident became more serious and the areas of ordered evacuations expanded, people were forced to evacuate and move around like this as a result. It is clear that hazards to health arose due to the inability of these people to respond to changes in their living environments.

Systems for Compensation

Japan's Act on Compensation for Nuclear Damage was enacted in 1961 (National Diet of Japan, 1961), eight years after Eisenhower's "Atoms for Peace" speech was given in 1953. Under this law, liability for nuclear damages was concentrated on carriers of nuclear power such as TEPCO, who came to bear unlimited liability. Meanwhile, the government is not held liable under this law, instead securing 120 billion yen in compensation for carriers of nuclear power from private insurance contracts.

For the Fukushima Daiichi Nuclear Power Plant accident, the Nuclear Damage Compensation and Decommissioning Facilitation Corporation (a government agency) has been providing support in compensation for TEPCO. As of 18 February 2022, the total amount of compensation paid by TEPCO is approximately 10.1957 trillion yen (Tokyo Electric Power Company Holdings, Incorporated, 2022), which greatly exceeds the amount of compensation from the government, with most of its financial resources composed of national taxes. Although the government is not liable itself, the money coming from taxes is paid through TEPCO, and it is unclear who bears responsibility.

Most of this compensation was paid to a limited number of Fukushima residents and businesses who resided in the evacuation order area near the Fukushima Daiichi Nuclear Power Plant. The claims ranged from mental and emotional damages, real estate compensation, housing security damages, incapacity to work damages, and business damages. In some cases, the amount ranged from tens of millions of yen to more than 100 million yen per household. On the other hand, many of the residents in urban cities, such as Iwaki City, Koriyama City, and Fukushima City, which have the largest populations in Fukushima Prefecture, were evacuated due to the nuclear accident, and since they had been by the government designated as "voluntary evacuation" areas and not areas for a compulsory evacuation order, only 120,000 yen per person was provided in compensation, and disparities in the amount caused conflict and discrimination among residents. For residents who had evacuated from the vicinity of the nuclear power plant, there was a risk of jealousy or prejudice if someone declared themselves as an evacuee, since such remarks mean a receipt of a large amount of compensation. And so, in many cases, people would stay home and avoid going out, their lives becoming inactive and their health deteriorating as they became unable to be active in the local community. Many nonprofit organizations (NPOs) and private organizations in Fukushima have been working to support the lives of residents at evacuation sites in order to eliminate this division between residents, but most of these organizations are small and were not able to carry out their activities continuously because they do not have ample funding to do so.

In the following, I explain the systems of compensation for the Fukushima Daiichi Nuclear Power Plant accident and the difficulties thereof. First of all, there are systems of providing condolence funds for disaster-related deaths by local governments. If a death is found to be disaster-related, 5 million yen is to be paid if the deceased was head of household and 2.5 million yen if not. Compared to compensation systems described later, such deaths are recognized with liberal "prima facie" explanations, but the longer the period of evacuation, the more difficult it becomes for them to be recognized as disaster-related deaths. In addition, in comparison with evacuees from areas of ordered evacuations, there is a tendency for very few cases to be recognized that involve residents who have evacuated from voluntary evacuation areas.

There are three main routes of compensation procedures that victims of a nuclear accident are able to take. The first is to make a claim with TEPCO directly. The victims themselves fill in the necessary items in TEPCO's written claim form and submit it with documents attached. Much of the compensation provided is done so through this procedure. However, as far as I know, there are more than 100 types of claims (mental and emotional distress, incapacity to work, real estate compensation, housing damages, business damages, home property and Buddhist altar compensation, etc.). The contents of the materials describing these are also written in fine print and are quite complicated. Even as lawyers, we have to put a lot of effort into understanding the details. It may have been impossible for many residents to read these materials, understand the contents, and carry out the procedures for making a claim. TEPCO staff contact victims to provide support and help them carry out the claim procedures. But for TEPCO itself, the entity responsible for compensation, to be assisting claimants in the first place is a logical contradiction. As far as I understand through legal counsel, there have been some cases here and there in which the procedures were completed with missing claims and an underestimation of compensation.

When there has been a case that can be regarded as a disaster-related death, staff from TEPCO would immediately meet with and interview the bereaved family. After requesting completion of a questionnaire titled "Confirmation of Daily Life", they gathered information that leaned toward a denial of a causal relationship with the nuclear accident, such as "illness was worsening before evacuation", "appropriate actions to maintain health were not taken during evacuation", and "suicide was committed due to family conflict". After a few months, a notice might be sent from TEPCO's attorney-at-law to the bereaved family, stating that there was no causal relationship with the nuclear accident. Even in cases that seem to involve typical disaster-related deaths, some procedures were completed with only a small amount of compensation paid, with TEPCO conclusively determining that the rate of impact from the nuclear accident is 15% without showing reasonable evidence to the bereaved families according to my interviews with them. However, as a result of the bereaved families in these cases performing the Alternative Dispute Resolution (ADR) claim procedure (which is described later), in cases in which no compensation had been granted, it was then granted, and in cases in which only a small amount of compensation was paid, the amount was significantly increased. To make use of the ignorance of the bereaved family of a deceased evacuee before they can consult with a lawyer about appropriate compensation, in order to discover information that is disadvantageous to the bereaved family and use it as reasoning for your counterargument to pay a small amount of compensation – I must say that such a procedure loses all fairness and neutrality.

The second method of achieving compensation is to use the Nuclear Damage Compensation Dispute Resolution Center, a form of ADR for the nuclear disaster. A system was established based on the fact that it would be extremely inconvenient for many evacuees for there to be no other procedures for relief other than legal trials after the Fukushima Daiichi Nuclear Power Plant incident. It is true that many evacuees were compensated through this system, but there was a major flaw.

This ADR is a "procedure for the intermediation of amicable settlement" between the victims and TEPCO and is not enforceable like trial and arbitration procedures. Since it is not a compulsory solution, from about 2015 – when public criticism of TEPCO's system had weakened – there has been a series of cases in which proposals for settlements presented by ADR mediators were rejected. For many collective ADRs, including one that filed 15,000 residents of Namie, TEPCO's repeated refusals to settle caused the ADR system for a quick resolution to fall into dysfunction – a major reflection on the systems for compensation concerning the accident at the Fukushima Daiichi Nuclear Power Plant.

There are also many difficulties in collecting data on deceased patients. In Japan, the legally required period for retaining medical records is five years, and in some cases, the medical records of the deceased had not been disclosed by the hospitals. In addition, even if an attempt was made to provide proof of the condition of one's health before the nuclear accident, in some cases, medical records at hospitals in the ordered evacuation areas had not been kept and could not be obtained. In order to carry out the procedure for compensation, it is necessary to obtain power of attorney from all heirs of the deceased, and even among cases I dealt with, I have had ten heirs living all over the country send me power of attorney. If there is even one heir who cannot be contacted, the procedure cannot move forward.

The third method is through court proceedings. Trials take at least three to five years. Many residents are people in their 60s, 70s, and 80s, and the situation is such that they do not have the financial strength and energy to keep TEPCO on trial for many years. Because there is not enough time remaining to aid these victims, it is necessary to design a system for quick relief so that residents do not have their suffering doubled by compensation procedures on top of a nuclear accident.

Listening to the Deceased

I was a victim of the Great Hanshin Earthquake in 1995. Of the 6,434 people killed in the earthquake, about 80% died from building collapses or fires, but within two months following the quake, 922 more were lost to disaster-related deaths. Sixteen years later, the Great East Japan Earthquake and the nuclear accident ended even more lives through disaster-related deaths.

Those who have been lost to disaster-related deaths could potentially have been saved by improvement of the environment at evacuation destinations, appropriate medical care, and help at evacuation destinations. We are able to hear from the victims who were lucky enough to survive disasters like major earthquakes and nuclear accidents, but we cannot hear the voices and messages of those who suffered the most: the dead. However, by listening to their bereaved families and reading their medical records, we can learn how they died and how they may have been able to live on. By doing so, we can make use of that knowledge for potential future disasters.

References

Japan Broadcasting Corporation (2019). *Genpatsu jiko: Inochi wo obiyakashita kokoro no kizu* [Nuclear accident: Trauma that threatened life] [Online]. Available at: https://www.nhk-ondemand.jp/goods/G2019095995SA000/ [Accessed: 21 August 2021]

Ministry of Economy, Trade and Industry (2012). *Tokyo Denryoku Gaisya no kyōkyū yakkan henko ninka shinsei ni kakaru satei hōshin* [Assessment policy for TEPCO's application for approval to change supply contracts] [Online]. Available at: https://warp.da.ndl.go.jp/info:ndljp/pid/8422823/www.meti.go.jp/press/2012/07/20120720002/20120720002-2.pdf [Accessed: 21 August 2021]

National Diet of Japan (1961). *Genshiryoku higai no hoshō ni kansuru hōritsu* [Act on Compensation for Nuclear Damage] [Online]. Available at: https://www.oecd-nea.org/law/legislation/japan-docs/Japan-Nuclear-Damage-Compensation-Act.pdf [Accessed: 21 August 2021]

Reconstruction Agency (2021). *Higashi Nihon Daishinsai ni okeru shinsai kanrenshi no shishasū (Reiwa 2 nen 9 gatsu 30 nichi genzai chōsa kekka)* [Number of deaths related to earthquakes in the Great East Japan Earthquake (Survey results as of 30 September 2020)] [Online]. Available at: https://www.reconstruction.go.jp/topics/main-cat2/sub-cat2-6/20201225_kanrenshi_teisei.pdf [Accessed: 21 August 2021]

Tokyo Electric Power Company Holdings, Incorporated (2017). *Fukushima Daiichi Genshiryoku Hatsudensho no genjyō to shuhenkankyō ni ataeru eikyō ni tsuite* [Current status of Fukushima Daiichi Nuclear Power Plant and its impact on the surrounding environment] [Online]. Available at: https://www.nsr.go.jp/data/000182828.pdf [Accessed: 21 August 2021]

Tokyo Electric Power Company Holdings, Incorporated (2022). *Baisyō kin no oshiharai zyōkyō* [Records of applications and payouts for compensation of nuclear damage] [Online]. Available at: https://www.tepco.co.jp/fukushima_hq/compensation/results/ [Accessed: 21 August 2020]

Part II
Living with Risk

6 Getting the Measure of Radiation Monitoring in Fukushima, Ten Years On

Louise Elstow

Introduction to Radiation Monitoring in Fukushima

The earthquake, tsunami and nuclear disaster that started to unfold in Japan in March 2011 – often referred to simply as the 'triple disaster' or '3.11' – marked the start of intense inspection of bodies and environments in the affected areas. Radioactive contamination from Fukushima Daiichi Nuclear Power Plant was dispersed across huge swathes of Fukushima Prefecture as well as several other neighbouring prefectures, including Ibaraki to the south, Tochigi to the south-west and Miyagi to the north (Hayakawa, 2012; Japan Ministry of the Environment [MoE], 2018). Thousands of people were evacuated or took the decision to evacuate from their homes; some have still not been able to or chosen to return (IAEA, 2015). In the days, weeks and, to some extent, even months after the disaster, little was known about where the contamination had been deposited, how much there was or what impact this might have on health long-term. Government information addressing this gap was slow to emerge. In those first months, acknowledging the paucity of publicly available Japanese government data, many citizens, activist groups and international organisations began to undertake their own measuring and monitoring, both to address a continuing dearth of granular data as well as to counter a lack of trust in the limited information being disseminated centrally (Kimura, 2016; Kenens et al., 2020; several of my own interviewees).

Ten years on, radiation measuring and monitoring remain common activities. They take place in multiple locations, from schools to service stations, and are conducted by numerous individuals, citizen groups, supermarkets, scientists and government bodies. Food monitoring stations continue to be found in all kinds of public and community spaces, while whole-body counters continue to monitor human bodies in hospitals, clinics and Citizen Radiation Monitoring Organisation (CRMO) offices. Various networks of people and devices exist to monitor, display and disseminate the information generated from these activities. The results are often made publicly available, typically in the form of tables, maps, charts or other visual representations that depict the radiological situation provided by the data. However, the simplicity of the numbers belies the complexity and multiplicity of generating them and interpreting them. I present the idea that radiation measuring and

DOI: 10.4324/9781003182665-8

monitoring are doing much more than merely identifying where radiation contamination is and how much there is, and set out a number of answers to the question, 'What is it for?'

This chapter draws on data from research I conducted into radiation measuring and monitoring after the Fukushima nuclear disaster. The findings are based on six months of ethnographic fieldwork in Japan, which was undertaken in two periods, the first in summer 2018 and the second in spring and summer of 2019. I conducted 35 semi-structured interviews with a diverse range of individuals engaged in radiation measuring and monitoring, as well as countless informal conversations in person throughout my Field trips and virtually since then. I spoke formally and informally with government agencies, local government officials, scientists, teachers, artists and affected citizens and citizen radiation monitoring groups (referred to henceforth as CRMOs following convention established by Kimura (2016)). The conversations were in Japanese or English, using a translator where necessary. I also undertook a significant amount of participant observation, including foraging for and measuring contaminated wild mountain vegetables, making my own radiation monitor, and accompanying scientists who were trying to determine why discrepancies exist between different sources of environmental data. In short, I tried to get under the skin of what radiation monitoring is actually about; what knowledges, devices and people it involves; why it is problematic; and why it remains important.

In this chapter, I invite readers to pay attention to what is going on behind the numbers produced by the network of government installed fixed radiation monitors. I want to answer the question, 'What is it for?' or rather, 'What is monitoring being used for?', the idea being that monitoring has been able to do things like helping reinstate important social practices or demonstrating care and attention and these things are not necessarily what might come to mind when considering what measuring and monitoring is there to do. I begin by exploring what exactly is meant by radiation measuring and monitoring before examining some of the reasons for doing it. I establish that radiation monitoring is a set of nuanced and problematic activities. The meaning of the data produced by radiation monitoring systems is open to interpretation, device failure and context. I then outline why and how radiation monitoring is different to radiation measurement and show that this difference is important but often overlooked. I then make the case for other kinds of work being done through radiation monitoring. My data suggest that radiation monitoring, whilst ostensibly concerned with providing answers to questions that address immediate health concerns, also extends to marking a place as special, enabling relatively mundane things like the sharing food with neighbours, selling produce or moving around different zones, as well as helping construct communities and denoting care and attentiveness.

Many different radiation monitoring infrastructures exist in Fukushima for monitoring human bodies, the air, the ocean, foodstuffs and more. In this chapter I have chosen to focus mainly although not exclusively on ambient radiation monitoring, in particular on the network of government installed fixed real-time radiation monitors visible all over Fukushima. They are typically found on the roadsides and outside other public places such as schools, government buildings, train stations and public parks, and are a unique and distinct example of Fukushima's many monitoring infrastructures. By February 2013, there were about 2,700 such monitoring posts in situ, a number which had swelled to 3,000 by 2018 (Kyodo, 2018; Nuclear Regulatory Agency [NRA], 2018). By following radiation monitoring devices and activities that they are part of, in a kind of 'follow the thing' multi-sited ethnography (Marcus, 1995), I hope to shed light on what is going on behind the data generated by them.

'What is it for?!'

In May 2019 I accompanied three scientists from the National Institute of Advanced Industrial Science and Technology (AIST) to Iitate Village in Fukushima Prefecture, north-west of the Fukushima Daiichi Nuclear Power Plant. The evacuation orders which had been put in place in 2011 had been lifted in Iitate in 2017, but only a small proportion of residents returned (McNeill and Matsumoto, 2017). Our visit was concerned with the network of fixed radiation monitoring posts around the village, which measure ambient dose rates, that is what dose a human body is being exposed to in that location. There were about 80 such radiation monitoring posts around the village, and a mixture of models was visible. Iitate village is a collection of twenty smaller hamlets that came together under one municipality in recent years and therefore covers quite a large area, about 230 square kilometres (ibid.). Therefore, the devices are not as densely distributed as the number might first suggest. The instruments installed by the national and prefectural governments from 2012 onwards were further augmented by a network of radiation monitors installed by the village authorities in 2016. The scientists informed me that Iitate is the only village in Fukushima to have done this. The village's monitoring posts provide additional points of data, alongside which data from the government installed posts can either be validated, contested or complemented. It is often possible to determine, just by looking at a device, whether it is a national, prefectural or local government monitor by the model used – each was a variation on a theme which includes a body housing the monitor itself, a solar panel and a display screen. This semi-standardisation suggests the potential for comparison across space and time between data from the various posts.

The intention behind our visit was for the scientists to scope out the monitoring posts' situations. They wanted to investigate discrepancies between the measurement data provided by the ground-based monitoring posts and a second set of data covering the same area but generated by aerial surveys. Land-based monitoring posts take a reading 1 metre above the ground (NRA, 2011), whereas aerial surveys conducted by helicopters monitor tracts of land between 300 and 600 m wide from the air at around 300 m above the ground. The equivalent reading at 1 m above the ground is then extrapolated from the aerial data (Miyahara et al., 2015: 16). Fixed monitoring posts therefore provide a smaller number of data points in static locations but are more locally specific. Aerial surveys smooth out local fluctuations by taking average measurements across a wider area. En route to the first monitoring post, one of the scientists explained that whilst there are clear rules and requirements for environmental radiation monitoring at designated nuclear sites, no such guidance or standards were available in other non-nuclear locations. He pointed out that this meant that there was no consistent approach to the siting and set up of the fixed radiation monitoring posts. The implications are that without an agreed standard for non-nuclear environmental monitoring, multiple inconsistencies have emerged in terms of ground preparation techniques, location choices and the amount of information provided to explain both the data shown and how it should be interpreted (Brown, 2012a, 2012b and 2015). The scientists' work was a first step in AIST's understanding of what kinds of variables might exist so that they could carry out a more detailed assessment at a later date and ultimately inform a future standard.

Partway through the day, we stood in front of a prefectural government monitoring post, the third such post of the day. It was adjacent to a road, behind it an intentionally waterlogged rice paddy awaiting the planting of the year's rice and opposite it over the road lay a farmhouse. The screen displayed 0.929 microsieverts per hour. Sieverts is a unit of the health effect of ionising radiation and government decontamination activities aim to reduce the level to 0.23 microsieverts per hour. One of the scientists used a Hitachi Aloka Survey Meter, a handheld device approved for government environmental radiation monitoring during decontamination works, to do a quick survey of the area just around the fixed monitor. Looking at the monitor from the road, the Aloka read 3.6 microsieverts per hour 1m to the right, 1.1 microsieverts per hour 1m to the left and on the road immediately to the front 0.3 microsieverts per hour. We were all unsure about why there was so much variation in such a small area, and this started to pose bigger questions.

A colleague of mine recently described the monitors as alien and out of place in farms like the one in Iitate. He appeared unsettled by the devices, as if they somehow interfered with his ideas of the Japanese rural landscapes. The two scientists and I discussed what the particular monitoring post we stood next to might mean for people living in the farmhouse opposite. The younger scientist suggested, 'The monitoring post is only correct *right* here. But people see it, those people in the farmhouse over there see it every day and think that it applies to all areas'. When I asked why that was a problem, the scientist paused before continuing: 'So the problem is … what *is* this?! What is it *for*? … [I]f people know that this monitoring post is unique, it is not a problem, but most people don't know'.

What the scientist seemed to be getting at was the situated (Haraway, 1988) and contextual nature of the information being produced by the monitoring post. The displayed number might be technically 'accurate' (e.g. the device was calibrated correctly and set up as per manufacturer's instructions), but that accuracy did not extend very far in a practical sense when being operationalized, and this was not clear to those viewing the screen. But his more general question is equally valid – what *is* it for? The 'it' in his question implicated not only the device in front of us but also the whole network of monitoring devices with digital displays set up around the prefecture. If this radiation monitoring post provides a number on its screen indicating the ambient dose rate, what is the purpose of this number? If there is no standard for siting the devices, then how does data generated by it link to the sets of data produced by other devices in the network? These monitoring posts routinely upload data into central records, and these data are then made publicly available – they could feasibly be used not just by the local residents to make life decisions but also prefectural, national or even international governments to make policy decisions. Bruno Latour suggests that scientific facts have to connect into a chain of facts that helps scientists' ideas travel between the local and the contextual to the universal (Latour, 1999). If scientific facts, such as the fact of radiation levels in this place as generated by this device, cannot be consistently translated along each step, then the chain of translations is not intact. So the local measurement lacks meaning at the universal level.

The scientist's observation calls into question also what it might mean if this number differs from the one shown on a post half a mile away, from a measurement produced by an aerial monitoring device or even an Aloka Survey Meter one step away from the post. What does it mean if the instruments start to fail? How long should such monitoring continue? When is enough data enough? How important is it to have a monitoring system in place nearly ten years after the initial disaster? After all, the farmhouse had been re-occupied, and surely its residents knew their local radiation levels by now.

Between Measuring and Monitoring

Before examining some answers to the question, 'What is it for?' I now make clear what I mean by *radiation monitoring* and *radiation measuring*. Often, the two terms are used relatively interchangeably or together as if describing a homogeneous activity. It is worth unpicking these terms because the monitoring posts are doing both measuring and monitoring; however, there are different implications for each.

Both radiation measuring and monitoring are activities involving various human *actors* (government agencies, scientists, activists, individuals, other non-activist groups, etc.) and non-human *actants* (devices, standards, methods, visual representations, etc.) that produce knowledges, understandings and ideas about radiation contamination – matter. This 'matter' then moves around the system to other individuals, via networks of paper maps, websites, spreadsheets, medical records, stickers and artwork, often translated along the way as it moves around. Measuring and monitoring infrastructures are inherently social and human things because the people living with them are influenced by the world produced by them (Jensen and Morita, 2017).

I take *radiation measuring* to be a one-off activity, whereby a device is used to generate, inscribe and present a reading in a standard unit of measurement. For example, fixed radiation monitoring posts display the most recent ambient dose rate measurement on the small screen at the top of the device. On the other hand, I use the term *radiation monitoring* to mean the ongoing and repeated measuring of a given item over a period of time. All monitoring involves an element of measurement; it also involves elements of comparison and repetition. Monitoring suggests a broader interest in trends over time and relationships between multiple data points and measurements. It speaks of persistence, both of the activity itself and of the contamination being monitored, as well as attentiveness and judgement.

I delineate between radiation measuring and monitoring because of the temporal and complexity differences between them. Measuring is useful in that it provides discrete points of data which may be of significance in and of themselves, but monitoring is a conceptually more intricate way of understanding radiological contamination. *Radiation monitoring* incorporates something more than merely the use of a device to measure radiation in an item at a given time in a given place. It provides over time, a build-up of an experiential baseline, formed from past and current conditions, from which future conditions might be judged. Monitoring helps identify changes, patterns and trends which may be of concern and involves paying attention to them. Undertaking monitoring indicates care and concern. It points to a potential future in which action might be taken to address the concerns identified. A single data point produced by the monitoring post we stood in front of might be enough to encourage a person or organisation to take a particular decision or action at a given time or with short-term implications (e.g. don't enter this place now, or don't eat this bunch of wild vegetables picked today), or the data might be part of longer-term monitoring over a longer period for that municipality or a greater geographic area. Monitoring is more likely to be linked with bigger decisions with longer-term impacts, in which case it is important to understand the contextual nature of the individual data points being produced.

Both measuring and monitoring, however, contain assumptions about what is to be measured and why. They are, for example, based on the consideration of generic standardised theoretical bodies that move and behave and are built in a particular way. These considerations have been distilled from lots of previous research and so they are intrinsically linked to a history of radiation monitoring that has determined the 'right' way by which to do this. Individual data points therefore never really stand alone as neutral entities.

In the following sections, I outline a number of possible responses to the scientist's question, *What is it for?* suggested by my data. In doing so I open up space for thinking about radiation monitoring in different ways and about the consequences of acknowledging the other work that it is doing.

Initial Questions of Safety and Later How to Live Again

At its heart, radiological measuring and monitoring are there to help an individual or organisation know the levels and locations of radiological activity. It is relatively easy to imagine that these activities are predominantly and consistently being carried out to answer questions about whether something is objectively safe or not. After all, much of the initial measuring and monitoring that took place in the weeks and months of the 3.11 disaster was to address the kinds of urgent issues, such as whether it is safe to stay here, which areas should be evacuated, where the worst of the contamination is (Blumenthal, 2021). An individual might ask, 'Should I leave? Is it safe to eat this food? Can we ask staff to work here?' All these questions relate to safety issues and those questions are of immediate relevance after an uncontrolled radiological release. Measuring and monitoring enable (and disable) people at an individual and organisational level to implement appropriate countermeasures and put in place radiation protection mitigations to address the risk suggested by the data – conventionally, then, monitoring and measuring universalise local data but put them into a broader context. In Fukushima, it was necessary to obtain information about contamination around the site very quickly, not least because the pre-existing network of fixed monitoring posts around the Daiichi site had been damaged by the earthquake and tsunami waves (Miyahara et al., 2015).

Over the years, these initial questions posed in the first days, weeks and months which focused on immediate safety have gradually been answered or rendered moot. Residents stuck in limbo have perhaps come to terms with living with the uncertainty of not knowing or perhaps they have moved on. In some cases, factors external to radiation may have overtaken evacuees' concerns about radiation safety – such as economic or family pressures to move back to the area (Weisshaupt, 2018). One of my interviewees who regularly goes back to Fukushima to undertake environmental radiation monitoring used the Japanese word *gaman*, 我慢, meaning 'patience' or 'endurance', when describing the situation in Fukushima. Given the volumes of data on radiation and contamination in Fukushima available these days, radiation measuring and its more attentive cousin, monitoring, seem to be doing more than answering questions concerning immediate safety and the aims, practices and performances of radiation measuring and monitoring are changing as time moves on and recovery from the incident is progressing.

A clinical psychologist who had worked in the parts of Fukushima affected by nuclear contamination for over six years offered one suggestion. He mentioned to me how the need for information had changed. In the beginning, residents and evacuees wanted to know whether they could return or not. Their requirements for scientific information were more general or conceptual. Now that the evacuation orders in many cases have been lifted, residents' concerns, he told me, were no longer about 'Sieverts vs Becquerels, they are about how to live. They are more specific'. They speak about radiation measurements in relation to questions such as 'Is it okay to eat this food that I have grown in my garden?' and 'Can I use the trees from my backyard as firewood as we would have done before the incident?' Essentially, they are grappling with whether it is possible to live their original lifestyle or what affordances they might need to make. The questions still relate to safety, but there is a more specific focus on a particular activity such as taking a family swim at a favourite beach in Iwaki City or producing a commercially viable bottle of Fukushima sake.

In this potential answer to the 'What is it for?' question, radiation monitoring allows individuals to identify specific ways of living with radiation and understanding the affordances that they might have to make. My data indicate that the initial period is marked by the expectation that government agencies will conduct measuring and monitoring at a macro level, which then leads to other exceptional activities taking place (evacuation orders, food controls, etc.), which are done on behalf of the affected population. Later on, in contrast, the sequencing starts with everyday practices that measuring or monitoring is then brought into and involves more individual and local focus. Obviously these two general statements are of course not universally true, as individuals will naturally have been concerned with their local situation in the earlier phases, but they would not reasonably be expected to have personal radiation monitoring equipment immediately to hand. Equally, government agencies continue in the later phases to be expected to produce radiation monitoring data and for that data to be publicly available. However, it is useful to note the broader pattern indicated, which moves radiation monitoring from being used in exceptional decision-making *on behalf of* citizens to everyday decision-making *by* citizens.

Unique Markers and Navigating New Ways of Knowing

As fixed parts of the built environment, monitoring posts pointed towards something unique and distinct about Fukushima. For me as a researcher, the existence of the fixed radiation monitors told me I was in the right place to research radiation.

Arriving in Fukushima for the first time, I remember seeing my first one – an oversized white bollard with a red LED screen and solar panel attached to a pole on the top. It was in the middle of the train station car park in Kōriyama City, a city which although not affected directly by the evacuation zones, still had levels of radiation elevated above pre-2011 background figures. As I left the station and saw Fukushima Prefecture for the first time, I must admit I was a bit excited, thrilled even, to finally be there and to spot a radiation monitoring post for myself. At last, I was able to see with my own eyes how much radiation there was in this potentially contaminated place I had heard so much about.

The human body makes for a poor radiation sensor. It is not able to detect the presence of radiation through the usual bodily senses such as sight, smell or taste. A large radiation monitor is a visible inscription and translation device, converting invisible radiation into visible numbers, to which I could then attach meaning about my surroundings, based on an understanding of what those numbers meant. The monitoring post was also a marker to me as an outsider, of what it meant to be contaminated. It meant I was in the right place; I had arrived in Fukushima and not, for example, Kyoto or Tokyo.

Over time, however, I learned a more quotidian way to note and pay attention to the numbers on the screens I saw and to the devices themselves. There was not so much excitement in these later encounters, as the frequency with which I saw the monitors rendered them into a familiar sight. However, I registered the radiation monitor number, much in the same way as I might note the price of fuel in petrol station forecourts in the United Kingdom – glance and move on. I started to know typical ambient dose levels in different parts of Fukushima Prefecture, to expect higher levels in Iitate or Yamakiya than I would do in Fukushima City or Suetsugi. I learned over time to pick up on larger fluctuations from these general baseline figures in each town and that residents in parts of the prefecture that typically had higher rates might tolerate a general level of radiation not considered acceptable elsewhere in the prefecture.

I also noted that some devices measured radiation in subtly different units: some were in units based on grays and some based on sieverts. Although in Fukushima's case, because of the radionuclides involved, a figure given in grays and a figure in sieverts, roughly amounted to the same number, having to navigate two different measurement units and in some cases display magnitudes was potentially confusing (a gray is a measure of the dose absorbed, while a sievert takes the type of exposure into account because some types are more damaging than others). I was told that some devices initially displayed measurements in nano-grays (the nano-gray is equal to one billionth of a gray – 10^{-9} Gy), which alarmed people by showing readings in triple digits, making them look much larger than the numbers shown on other devices measuring in microgray per hour μGy/hr or microsievert per hour μSv/hr. To address this alarm, the measurement units painted on the side of the screen were taped over and the device was reset to μGy/h, thus moving the decimal place over to a less concerning location. Elsewhere I was shown pictures of small manufacturer's screens in nano-grays, augmented by larger screens displaying the same amount in μGy/h.

Furthermore, I noted that, as in Iitate, whilst the number on the screen might be technically accurate for that location, the accuracy might only extend a short and unspecified distance away. For example, the reading from a monitoring post in the centre of the car park related only to the dose rate for that specific spot. It might not reflect the situation in the whole city or even perhaps the whole car park.

On various occasions, I held different handheld devices next to the fixed radiation monitors and saw a degree of disparity in the readings. This was a standard occurrence according to a number of people I spoke to – professional and citizen-scientists, as well as government officials. It was up for debate as to which of the devices was 'right' and which were inaccurate. There were contrasting views on how to judge accuracy: whether this should be judged on the number of data points, the time elapsed since the last measurement, the people doing the monitoring, the device they used or the method of using the device. There is a rich array of approaches to, motivations for and results emerging from radiation measuring and radiation monitoring activities.

Allowing Discussion, Validating Concerns and Other Ways of Knowing

In August 2018, I spoke with Yuko-San (pseudonyms are used throughout this chapter to maintain anonymity) about some of the measuring and monitoring she had carried out in the previous seven years. Soon after the disaster, feeling lonely and frustrated at a perceived lack of concern by other mothers about the information being provided by the government, she decided she wanted to do something more proactive. In May 2011, she joined a volunteer group in her town, working on supporting those affected by the disaster. She asked to be part of the radiation monitoring team and was given a handheld radiation monitor and asked to measure and record the ambient dose in various locations three times a day. She was already the leader of a separate volunteer group from before the incident and offered free radiation measurements to her existing network. When doing house visits to monitor radiation, she found would spend a lot more time speaking to residents about radiation than she did taking measurements. Discussing the measuring activity allowed for dialogue about radiation concerns, a subject she maintained was hard for families to discuss openly.

Later, Yuko became involved in another monitoring group specifically concerned with radiation hazards in schools and proactively campaigned for changes to food in schools as well as decontamination practices. Government-installed fixed radiation monitoring posts have been prominently positioned in Fukushima's schoolyards since 2012. Schools were also measured in a different way by government decontamination workers prior to and after decontamination work, to determine how much decontamination to undertake and whether the activity had been successful. This measuring process included taking one measurement in the centre of the playground and one in each of the corners, up to five measurements in total (Ministry of the Environment, 2013: 1–10). The mean dose rate was calculated by averaging out the measured values, which then determined if any further decontamination was needed. The aim was to reduce the average ambient dose rates down to 0.23 microsiverts per hour. I spoke to several organisations and individuals who were not satisfied with this method. They were concerned that children were being subjected to levels of radiation not reflected by the data being produced by government monitoring activities and that localised hot spots were being missed. Two CRMOs I spoke to (one of which was Yuko-San's) had carried out additional monitoring in Fukushima's schools: one monitored inside classrooms and buildings using a device that sucked in air and particles; the second focused on the rates outside in the school grounds using portable monitoring devices.

The second CRMO described how, in 2013, it was successful in getting permission from the local education committee to undertake detailed monitoring in school grounds in their city. The group's approach to monitoring differed from the fixed monitoring post system and that of the government decontamination teams. Not only did the CRMO take more measurements, but it also monitored in more places and at different heights. Its teams surveyed all over the school grounds, taking in some cases more than 100 measurements and paying particular attention to hot spot areas. Government monitoring is normally taken at a height of 1 m; however, the Decontamination Guidelines allow for readings to be taken at 50 cm above the ground in elementary schools. The CRMO measured at a height of 50 cm, but if its members found that the reading was higher than 0.15 microsiverts per hour, they would take another measurement at ground level. The group also took soil samples at those locations and sent them to a lab in Tokyo for testing. Because there is no standard for soil contamination that determines whether an area is decontaminated further, the CRMO uses the standard set for food to determine whether it should ask for further decontamination to take place following its survey. In lieu of an official standard, they used the food standard because children might end up ingesting the soil during play. So, if a soil sample was above 100 becquerels per kilogram, or the air dose rate above 0.15 microsiverts per hour, the CRMO made extensive pleas for further decontamination. The becquerel is another unit of measurement relating to radioactivity; however, this unit relates more to the element decaying rather than what the radiation might affect, such as the effect on the human body, which is inferred by a measurement in sieverts. One becquerel is one decay per second. Food is measured in this way, and most foods in Japan have a limit of 100 Bq/Kg. Therefore, the use of both units of measurement means that the two measurements are not directly comparable.

Whilst the government decontamination guidelines seem to warn those taking measurements to avoid places where there might be concentrations, such as 'depressed areas and puddles, under rainwater guttering … or near trees' (Ministry of the Environment, 2013: 19) the group that I spoke to actively sought some of these places out. This was not because of a misguided or biased hot spot-hunting tendency. Their reasoning was that places like gullies down the side of buildings or areas tucked in behind a shed are exactly the kinds of places that children like to go and hide or play in. Children are more likely to seek these places out. This example suggests that whilst adults might be assumed to want logically to avoid contamination, the same may not be said about a child, who may seek to explore and move around the world in a different way. In March 2015, the group monitored a school site and identified a patch of soil which measured 130,000 becquerels per kilogram, 1300 times the limit for food. The government decontamination efforts had missed this area during their monitoring; it was not recognised by government workers as a place where children would go, worthy of monitoring.

Many CRMO members were mothers concerned about the health of their children. Aya Kimura (2016), in her work on the gendered politics of citizen science, argued that citizens engaging in radiation measurement and monitoring is part of the stand against the gendered policing of radiation concerns. Women felt that their concerns about potential contamination were seen as anti-science and irrational. Since 2011 the term *fūhyōhigai* (harmful rumours) has been used frequently in Japan in relation to Fukushima, and even by the Japanese Prime Minister (Kimura, 2019) to denote supposedly unfounded concerns about radiation contamination (Brasor, 2011; Yamaguchi et al., 2016; Stolz, 2018). It has been blamed for causing immense economic damage, creating unnecessary food avoidance and inflicting unnecessary harm and suffering on those in the affected areas and was raised frequently in many of my conversations not only with CRMOs but with doctors, lawyers and classically trained scientists as well. Mothers who raised concerns about radiation in schools were often accused of *fūhyōhigai*. Kimura suggests that measurement and monitoring undertaken by women in citizen science groups was a way to counter these claims of *fūhyōhigai* by grounding concerns in science and making them valid again. Yuko was aware of not wanting to be seen as an irrational mother or 'monster parent' when raising her concerns about contamination at schools. She said she wanted to get rid of the traditional image of the mother being hysterical and emotional, so she tried to be really calm, speak slowly and behave in a really 'cool' way, using science to get her point across.

In this example, there are several responses to 'What is it for?' First of all, measuring was a way for Yuko to take a proactive step towards helping after the disaster as well as finding a way of addressing her own concerns about inadequate government information. Later on, when involved in the school monitoring project, Yuko used monitoring to put pressure on the local government to undertake more or more focused decontamination in areas identified as a concern, but it was also a way of making her voice heard when it otherwise might not have been. Using the science of monitoring to back up her claims validated them and made them less easily dismissed.

(Re)building Communities

As well as providing an additional point or counter-point to the official government radiation measurements, community-level ambient radiation monitoring can also be about building and rebuilding communities. Detailed radiation maps were not produced by the government for months. Because of the delays and challenges in getting data from official sources, one outspoken scientist, who produced several well-known maps of Fukushima contamination initially via Twitter, observed that 'most people lost trust in the scientist of the governmental side, so they did not trust these data directly, so they wanted to check the contamination themselves'. That is not to say that the government data were always distrusted but that checking contamination themselves allowed citizens to feel more confident in the information they were being provided by the government. It was, in a way, a form of data calibration.

One evening in a typical Japanese style inn in a village close to Fukushima Daiichi, I spoke with the owner, Tanigawa-San. Alongside running the inn, he and his wife are part of a local group with links to a group based in Chernobyl, who produce their own local radiation data and radiation maps. The couple were some of the earliest returnees to the village. For years now, they have been encouraging people back to the area and are also helping promote innovative farming projects that foster links between local farmers and international businesses.

Tanigawa-San explained to me that the audience for the maps is the people who live here and the people who want to return. The maps cover four municipalities close to the power plant; in some cases, they gained permission from the local mayor to carry out the monitoring. The area is divided up into a mesh of squares, each 500 m² square assigned a colour depending on the level of radiation measured. If the quadrant is inaccessible, because for example it covers forest or mountains, the base map is visible, indicating that radiation monitoring is not possible. This format was translated from similar maps produced by community groups after Chernobyl with whom the group have established links, according to the monitoring station's website. Measurements are taken by 14 teams, with each team responsible for 30 squares on the map. The process is relatively manual, in that the figure is read off a screen and then input by hand onto a data sheet, which is then translated into coloured squares on the map. Each paper map is, in fact, two maps, one showing the most recent results and one showing the previous round of results. By 2019, the group had produced these maps 16 times, roughly every six months since July 2011, although recently the group decided to reduce this to just once a year. Measurements are taken in the same single location every time. The paper maps are printed off and made available via the group's main office in a town about 20 km away, as well as local publicly accessible places like the inn itself, shops, banks, schools and the city office.

Another CRMO offered to help Tanigawa-San's group make recording the data easier and less time-consuming by supplying a device that records the data and location automatically for download at a later point. Another research institute showed the group how to disseminate its data more widely using digital copies online. Both suggestions appear to have been politely declined or quietly ignored. This quiet dismissal intimates to me that the purpose of Tanigawa-San's group monitoring activities is now less about the data values it produces and more about bringing together people from local communities. As a result, maps and data are not able to travel far or to be used by anyone outside the community.

I sensed the value of the group's current monitoring activities no longer lay in establishing where was safe or where was not. Monitoring was now one of several mechanisms (along with the farming project and encouraging visitors back to the inn) for bringing the community together in a sustained manner to demonstrate the possibility of re-establishing a life in the village. In this example, measuring and monitoring are positioned in relation to community building and trust. Together, the various mechanisms showed others who had still not returned to their former homes that radiation levels continued to decrease, that the area was habitable and that living is an economic and social possibility.

Indicating Improvement, Stability and Decline

One interviewee opined that when first installed by the government, fixed radiation monitoring posts 'show[ed] that somebody is paying attention. Which was the big problem before, that nobody was paying attention'. Their installation marked the start of more visible routinised monitoring by the government, indicating that they were not only taking action to deal with the disaster but also doing so in an ostensibly transparent fashion – the data were visible for anyone to see.

However, over the course of two days in April 2019, I passed four or five fixed radiation monitoring posts showing large, obviously erroneous numbers on the screens or whose screens were no longer facing the road. A prefectural government official later explained it was probably because they were broken and that this was becoming quite common. Designed with a ten-year life span, they were not intended to last beyond 2022. One of the main contaminants released in the disaster was radioactive iodine 137 which has a radioactive half-life of about 30 years. Therefore, it seems an odd choice to design a network that is going to start to fail in a third of the time. Given that the NRA argued that the maintenance bill is now an unpalatable 360 million yen or more (3.5 million US dollars at current exchange rates), the sight of failing radiation monitoring posts might become increasingly common (NRA, 2018).

In 2018, the prefecture proposed to remove or re-site some monitoring posts. Some posts would be removed entirely; others would be re-sited elsewhere in the prefecture, closer to the nuclear power plant (NRA, 2018). At the time the NRA argued that 'continuous measuring is unnecessary in areas where dose rates are low and stable' (Higashiya, 2018). Elsewhere, the maintenance costs were cited (Kyodo, 2018). The NRA therefore view the function of the monitoring post as purely to demonstrate that levels of radiation are 'low and stable'; monitors become an unnecessary expense. In this example, the answer to the question, 'What is it for?' is positioned in relation to budgetary and financial constraints. The current monitoring relates to a previous radiological incident, in this case the 2011 contamination, is always at risk of budgetary removal. It must fight to be permitted to relate to any potential future incidents resulting in elevated radiation levels (e.g. another accident at Daiichi during decommissioning or recontamination from existing contamination movement).

The NRA's position is not reflected in the views of some citizens or local authorities, however (Ishizuka and Tomatsu, 2018). A priest stressed to me that his congregation viewed monitoring posts as 'lucky charms' and that residents were not concerned by them. He described the network as a safety blanket and maintained that the monitors made residents feel safe. When consulted on the proposed changes, many municipalities (NRA, 2018) indicated that they wanted monitors to remain in place, particularly in schools and parks. They argued that contaminated waste sites around the prefecture provided a potential source of future contamination, as did the ongoing decommissioning work at the damaged nuclear site. Monitors were needed to track any such upward trends. Furthermore, some organisations commented that posts reduced public anxiety and that fixed monitoring posts would need to be replaced by another publicly accessible monitoring system. Some citizens petitioned for the proposed plans to be scrapped.

Some municipalities, however, did support the proposed plans. These tended to be those further away from the plant in places like Aizu, a mountainous area to the far west of the prefecture. Radiation levels in Aizu had been consistently low since the disaster, and the monitors were daily reminders of the incident or contamination that had not particularly affected them. According to one Fukushima Prefecture official I spoke to about the plans, some communities viewed the disaster as something that happened to other people, not them; therefore, the monitoring posts were an unwanted and damaging reminder of someone else's incident, in some ways a visual form of *fūhyōhigai*.

On returning to Fukushima in 2019, I was told that, following public outcry and bad press, the plan to change the monitoring posts was not implemented; however, if failures cease to be repaired then one of the most common visible signifiers of potential contamination is becoming gradually eroded nonetheless. In this last account, the monitors are at once both a sign of care and attention by the government, as well as an indicator of that care and attention dwindling. Whilst the focus of some residents and organisations is on the current and potential future contamination risk, the changes proposed by the prefectural government suggest that they would like to slowly begin the process of decoupling the present less contaminated Fukushima from the events of 2011.

Conclusion

In this chapter, I invited readers to start to think about what radiation measuring and monitoring are for and what they are actually doing. Using environmental radiation monitoring as a starting point, I use Fukushima's fixed radiation monitor network and selected examples from CRMO environmental monitoring practices to demonstrate that measuring and monitoring activities generate data that are open to interpretation, are contested and are multiple. I set out a clear delineation between the relatively simple act of monitoring, in contrast to the more complex notion of radiation monitoring. Monitoring is shown to involve not only devices, methods, actors and standards but also time and attentiveness. I argue that the different ways that ambient measuring and monitoring are carried out by different individuals, groups and institutions construct different and highly contextual views about the levels, locations and duration of the contamination in Fukushima, as well as tensions between the right and the wrong way of doing it. To a degree this is inevitable, and the ongoing challenge is to articulate what is happening within different groups so that these differing methodologies and practices can work together productively rather than in conflict.

Eleven years after the disaster, both radiation measuring and monitoring continue to be important, if contested, activities but their importance extends beyond merely knowing where contamination is and how much there is. Initial monitoring expectations focus on the expectation that government agencies will conduct measuring and monitoring at a macro level in order to be able to tackle big-ticket questions for the purposes of population-wide radiation protection measures on behalf of the affected population. Later on, in contrast, the sequencing tends to start with everyday practices that measuring or monitoring is then brought into and involves more individual and local focus. The broader pattern indicated moves radiation monitoring from being used in exceptional decision-making *on behalf of* citizens, to everyday decision-making *by* citizens.

My data have established several possible answers to the scientist's question – 'What is it for?' Measuring and monitoring activities can signify care and attention towards a population – merely the act of monitoring one thing rather than another points towards the measured item being of importance. Measuring and monitoring can link what is happening now to a past or a future toxic event and is part of an ongoing attentiveness associated with a living with uncertainty. Measuring and monitoring allow exceptional and mundane decisions to be taken. Measuring and monitoring can enable a claim for accuracy or give a voice to or validate a concern which might otherwise be dismissed. Whilst radiation measuring and monitoring, of course, overtly continue to be about the implications of radiation on human health, the work that they are doing does not end there.

References

Blumenthal, D. (2021) *Measurement and data from citizen science devices, roundtable discussion*. SAFECAST 10 – Live from Fukushima, Around the Globe, 10th Anniversary Event, 13 March 2021, Online.

Brasor, P. (2011) Japanese officials dress the part but fail to address the issues. *Japan Times*, 27 March 2011 [Online]. Available at: https://www.japantimes.co.jp/news/2011/03/27/national/media-national/japanese-officials-dress-the-part-but-fail-to-address-the-issues/ [Accessed: 11 January 2021].

Brown, A. (2012a) Information, misinformation, disinformation (or, "These aren't the droids you're looking for"), Part 1. *Safecast Blog*. 29 December 2012 [Online]. Available at: https://blog.safecast.org/2012/12/information-misinformation-disinformation-or-these-arent-the-droids-youre-looking-for-part-1/ [Accessed: 31 January 2021].

Brown, A. (2012b) Information, misinformation, disinformation (or, "These aren't the droids you're looking for"), Part 2. *Safecast Blog*. 29 December 2012 [Online]. Available at: https://blog.safecast.org/2012/12/information-misinformation-disinformation-or-these-arent-the-droids-youre-looking-for-part-2/ [Accessed: 31 January 2021].

Brown, A. (2015) More droid trouble. *Safecast Blog*. 9 May 2015 [Online]. Available at: https://blog.safecast.org/2015/05/more-droid-trouble/ [Accessed: 31 January 2021].

Haraway, D. (1988) Situated knowledges: The science question in feminism and the privilege of partial perspective. *Feminist Studies*, Vol. 14 (3), p. 575–599.

Hayakawa, Y. (cartographer) (2012) *Radiation contour map of the Fukushima Daiichi accident*. 7th ed [Online]. 8 August 2012. Available at: http://kipuka.blog70.fc2.com/blog-entry-535.html [Accessed: 5 August 2021].

Higashiya, M. (2018) NRA to remove most dosimeters in Fukushima as radiation drops. *The Asahi Shimbun*, 5 April 2018 [Online]. Available at: http://www.asahi.com/ajw/articles/AJ201804050009.html [Accessed: 4 January 2019].

IAEA (2015) *The Fukushima Daiichi Accident: Report by the Director General*. Vienna: IAEA.

Ishizuka, H. and Tomatsu, Y. (2018) Locals opposed to removal of most dosimeters in Fukushima. *The Asahi Shimbun*, 9 July 2018 [Online]. Available at: http://www.asahi.com/ajw/articles/AJ201807090004.html [Accessed: 4 January 2019].

Jensen, C.B. and Morita, A. (2017) Introduction: Infrastructures as ontological experiments. *Ethnos*, Vol. 82 (4), pp. 615–626.

Kenens, J., Van Oudheusden, M., Yoshizawa, G. and Van Hoyweghen, I. (2020) Science by, with and for citizens: Rethinking 'citizen science' after the 2011 Fukushima disaster. *Palgrave Communications*, Vol. 6 (1), pp. 1–8.

Kimura, A. (2016) *Radiation Brain Moms and Citizen Scientists: The Gender Politics of Food Contamination After the Fukushima*. Durham: Duke University Press.

Kyodo, (2018) Radiation monitors in Fukushima to be scrapped after malfunctioning to the tune of ¥500 million a year. *Japan Times*, 21 May 2018 [Online]. Available at: https://www.japantimes.co.jp/news/2018/05/21/national/fukushima-prefecture-radiation-monitoring-posts-installed-3-11-hit-glitches/ [Accessed: 31 January 2021].

Latour, B. (1999) Circulating reference: sampling the soil in the Amazon forest. In Latour, B. *Pandora's Hope: Essays on the Reality of Science Studies*. Cambridge, Mass.: Harvard University Press, p.24–79.

Marcus, G. (1995) Ethnography in/of the world system: The emergence of multi-sited ethnography. *Annual Review of Anthropology*, Vol. 24 (1), pp. 95–117.

McNeill, D. and Matsumoto, C. (2017) In Fukushima, a land where few return. *Japan Times* [Online]. Available at: https://www.japantimes.co.jp/news/2017/05/13/national/social-issues/fukushima-land-return/ [Accessed: 31 January 2021].

Ministry of the Environment (2013) *Decontamination Guidelines*. 2nd ed. [Online]. Available at: http://josen.env.go.jp/en/policy_document/pdf/decontamination_guidelines_2nd.pdf [Accessed: 31 January 2021]. (Tentative translation).

Ministry of the Environment (2018) *Result in the intensive contamination survey area* [Online]. Available at: http://josen.env.go.jp/en/decontamination/ [Accessed: 31 January 2021].

Miyahara, K., McKinley, I., Saito, K., Hardie, S. and Iijima, K. (2015) Use of knowledge and experience gained from the Fukushima Daiichi Nuclear Power Station accident to establish the technical basis for strategic off-site response. *JAEA Review*, Vol. 2015-001.

Nuclear Regulation Authority (NRA) (2011) *Comprehensive radiation monitoring plan* [Online]. Available at: http://radioactivity.nsr.go.jp/en/list/273/list-1.html [Accessed: 30 December 2018].

Nuclear Regulation Authority (NRA) (2018) *Review of placement of real-time dosimetry system* [Online]. 20 March 2018. Available at: https://www.nsr.go.jp/data/000224268.pdf [Accessed: 5 August 2021]. (In Japanese).

Stolz, R. (2018) Money and mercury: Environmental pollution and the limits of Japanese postwar democracy. *Positions: Asia Critique*, Vol. 26 (2), pp. 243–263.

Yamaguchi, T. (2016) Scientification and social control: Defining radiation contamination in food and farms. *Science, Technology and Society*, Vol. 21 (1), pp. 66–87.

Weisshaupt, M. (2018) From self-evacuees to returnees. *International Symposium: Fukushima Nuclear Evacuees: Researchers' Findings and the Voices of the Victims*, 7–8 July 2018, Tokyo: Meigu Gakuin University.

7 Prioritising Health Risks after the 3.11 Disaster

The Application of Wellbeing Indicators

Michio Murakami

In the aftermath of the 3.11 disaster, not only radiation but also secondary health effects such as increases in lifestyle-related diseases, psychological distress and a decline in wellbeing occurred. How should policies prioritise these various risk issues? This chapter introduces the methodology and case studies of comparing risks and discusses issue-setting, based on citizen participation and interviews with risk communicators working with citizens. Finally, a case study of research analysed to improve wellbeing after the 3.11 disaster is presented.

What Is 'Risk Research'?

There are many definitions of *risk*. The word originates from the Latin *risicáre* or Italian *risicare*, which means to navigate among cliffs (Kanoshima, 2000; Bernstein, 1996). Originally, the concept was active, with blame placed on the actor. Currently, two definitions set out by the International Organization for Standardization are well known in the field of risk science: a 'combination of the probability of occurrence of *harm* and the severity of that harm' (ISO/IEC, 2014: 2) and an 'effect of uncertainty on objectives' (ISO, 2009). In the *Encyclopaedia of Risk Research*, edited by the Society for Risk Analysis Japan (2019), the common aspect to the various concepts of risk is the possibility that a cause or event may cause something unfavourable to happen to the protected object. For example, the magnitude and frequency of hazards and threats change with advances in technology and society (e.g. climate change). However, the scope of protection is also influenced by society's changing values and can be expanded from human health to ecosystems, human rights and personal information. Risk, therefore, is a concept that is always closely connected to society and values.

Risk research is a collection of diverse disciplines concerned with individual and social decision making around the handling of risk (Society for Risk Analysis Japan, 2019). One of the major roles of risk research is to provide scientific knowledge and tools for social implementation to support decision-making about what actions or measures are appropriate. One of the most important aspects of risk research is that each action or measure is associated with multiple risk events, resulting in trade-offs (Graham and Wiener, 1997).

DOI: 10.4324/9781003182665-9

For example, while treatment of raw tap water produces disinfection by-products, it is also beneficial because it disinfects pathogenic microorganisms, greatly reducing the risk of infectious diseases from the use of tap water. By assessing the magnitude of such diverse and heterogeneous risks, risk research supports decision-making regarding measures and actions. Science that separates purpose and value from the activity of knowledge and explores nature from a purely objective standpoint (science for science's sake) is 'cognitive science', whereas science that aims to realise purpose and value and promotes overarching research (science for society's sake) is 'design science' (CNAS, 2003). Risk research focuses on design science. In this chapter, in addition to describing the significance, methodology and examples of risk comparisons and indicators (Murakami, 2019), I discuss approaches for determining which risks should be resolved, as well as how to evaluate policies aimed at improving wellbeing.

Multiple Health Risks after a Nuclear Disaster

The Fukushima Daiichi Nuclear Power Station disaster in March 2011 resulted in limited radiation exposure due to the prompt implementation of measures such as evacuation and food shipment restrictions (UNSCEAR, 2014; IAEA, 2015). However, it exerted secondary health effects other than radiation exposure. One of the typical secondary health effects that occurred during the acute phase was the increase in mortality associated with the evacuation of elderly people from nursing homes. In five nursing home facilities in Minamisōma City, the mortality rate of residents increased by a factor of 2.68 for almost a year immediately after evacuation (Nomura et al., 2013). This was not due to deaths caused by the earthquake itself or exposure to radiation but rather due to the physical and mental burden of the evacuation and limited medical resources.

In the chronic phase, many of those affected by the disaster suffered physical and mental health effects. In Minamisōma and Soma Cities, the age-adjusted prevalence of diabetes in 2014 increased to 1.6 times the pre-disaster level (Nomura et al., 2016). This is not due to the effects of exposure itself but to lifestyle and environmental changes associated with evacuation. Furthermore, approximately one year after the disaster, the number of people showing psychological distress increased to nearly five times that of normal times and, although it has since gradually decreased, it is still higher than normal (Maeda and Oe, 2017). In addition, a decline in wellbeing, such as life satisfaction and positive emotions, has been observed in the population since the disaster (Hommerich, 2012; Rehdanz et al., 2015; Tiefenbach and Kohlbacher, 2015; Murakami et al., 2017a). More on wellbeing and happiness is discussed later in this chapter. This decline is especially severe for those who cannot decide whether to return to their hometowns after evacuation (Murakami et al., 2020a). Other impacts include family separation, economic impacts, unemployment, radiation anxiety, stigma, prejudice, community fragmentation, and increased need for nursing care (Sawano et al., 2018; Sone et al.,

2016; Hasegawa et al., 2019). Thus, the impacts of a nuclear disaster not only are limited to radiation exposure and physical and psychological health effects but also extend to social health (i.e. the social determinants of health), including family, community and employment.

How should society respond to such diverse issues as various types of diseases and declining wellbeing after the 3.11 disaster? Since time, budget and possible measures are finite, there is a need to prioritise in order to reduce risk. Given the wide variety of health risks, however, it is not obvious which risk issues should be prioritised and priorities can differ between various stakeholders. A promising approach is to compare multiple risks using the same index. By comparing the magnitude of risks quantitatively, using the same indicators, we can learn which risk issues are important and worth reducing. Of course, it is not sufficient to discuss only the size of the risk or cost-effectiveness. From a utilitarian perspective, such calculations may result in the improper conclusion that measures need not be taken to address rare diseases because they are not cost-effective. My intention here is to illustrate that comparing various risks is necessary and can provide much-needed information for societal discussion and policymaking.

Indicators for Risk Comparison

In assessing risk, the question arises as to which indicators should be used. This question is not trivial, because the choice of indicators impacts on subsequent resource allocation and policy decisions. One of the most common indicators is mortality rate. There seems to be an implicit social consensus that a low mortality rate is a positive thing. The earliest historical example of risk comparison using mortality rate (more accurately, the proportion of deaths from the disease to the total number of death) dates to the mid-seventeenth century when John Graunt (1662) published a book in which he argued that we can understand which diseases are feared by comparing the number of deaths in England from different causes. However, there are aspects of mortality that are not reflected in mortality rates, such as the timing of death. For example, the timing of cancer deaths due to radiation exposure and deaths due to evacuation from nursing homes are completely different. On average, cancer deaths due to radiation exposure are estimated to peak several decades after exposure, whereas deaths due to evacuation from nursing homes often occur within the first three months.

An alternative indicator is life expectancy. A loss of life expectancy represents the average length of shortened life expectancy in the population as a whole, which can account for differences in the timing of deaths. A loss of life expectancy is based on the idea that a day is worth the same to everyone, and two ethical justifications have been argued for it: health-maximising ageism (Tsuchiya et al., 2003) and 'fair innings' ageism (Williams, 1997). Health-maximising ageism is concerned with the efficiency perspective of maximising people's total life expectancy, whereas 'fair innings' ageism is concerned with the equity perspective that humans have the right to enjoy the same number of years of life. The concept of loss of life expectancy often leads to debates about whose life is more important: the life of the young or of the old. However, from an individual's perspective, if there is, for example, a risk event A with a mortality rate of 10% at age 20 and a risk event B with a mortality rate of 30% at age 80, it is possible to determine that the risk of B is less than that of A. This is because we are focusing on life expectancy, rather than the magnitude of the mortality rate itself. Loss of life expectancy has been used to compare the magnitude of various risks (Cohen and Lee, 1979; Gamo et al., 2003). In the field of radiation protection, it has also been used to discuss the validity of setting regulatory values (ICRP, 1991).

Indices such as Quality-Adjusted Life Years (QALYs) and Disability-Adjusted Life Years (DALYs) have been developed and used to compare risks by proposing methodologies to weigh the decline in quality of life, instead of focusing only on death (Zeckhauser and Shepard, 1976; Murray et al., 1994). In particular, DALYs have been used to compare the magnitude of various risks globally and over time to assess the priority and effectiveness of measures (GBD 2015; DALYs and HALE collaborators, 2015; Forouzanfar et al., 2016).

Furthermore, there are indicators that assess quality of life in terms of wellbeing (Johnson et al., 2016). My co-authors and I have developed an indicator of happy life expectancy and loss of happy life expectancy (or gain of happy life expectancy; Murakami et al., 2018; Arai et al., 2020; Murakami et al., 2021). Happy life expectancy is the life span that people live with a subjective emotional feeling of wellbeing. A loss of happy life expectancy is the decrease in happy life expectancy due to a risk event; a gain in happy life expectancy is the increase in happy life expectancy. For this indicator, a decrease or increase in quality of life is assessed using levels of emotional wellbeing. This was developed to account for the importance of building a society happier than even before the 3.11 disaster, in addition to accounting for longer life expectancy. Happy life expectancy, which concerns positive emotions, is measured using a life table and a questionnaire: 'Did you experience a feeling of happiness yesterday? [Yes or No]' (See Table 7.1 and later discussion for more on this). This index was developed in response to the fact that happy life expectancy is an accumulation of momentary feelings of emotional happiness, in order to allow the measure to be capable of arithmetical calculation.

A characteristic of this index is that it can evaluate not only illness and disability but also the acquisition of happiness through, for example, living with family or living in one's hometown. In fact, the life expectancy gained by returning to one's hometown after the 3.11 disaster has been evaluated using this index (Murakami et al., 2021). The concept of happy life expectancy can also lead to debates about whose life is more important – the young or the old. For example, from an individual's perspective, if Life A is happy at birth but gradually declines in happiness and is unhappy before death and Life B is less happy at birth but gradually increases in happiness and is happy before death, most people would deem Life B as a better life (Kagan, 2012). This can be explained by the peak-end rule, where cognition of an event is determined

Table 7.1 Forms of wellbeing and corresponding questionnaire items (Murakami et al., 2020a)

Type of wellbeing	Subcategories	Elements	Questionnaire items
Momentary feelings of emotion	Positive emotion	Enjoyment	Did you experience a feeling of enjoyment yesterday? [Yes or No]
		Emotional happiness	Did you experience a feeling of happiness yesterday? [Yes or No]
		Laughter	Did you laugh yesterday? [Yes or No]
	Negative free emotion	Stress free	Did you experience a feeling of stress yesterday? [Yes or No] (reverse score)
		Sadness free	Did you experience a feeling of sadness yesterday? [Yes or No] (reverse score)
		Worry free	Did you experience a feeling of worry yesterday? [Yes or No] (reverse score)
Judgements about feelings over the long term	Life satisfaction and general happiness	Life satisfaction	All things considered, how satisfied are you with your life as a whole these days? [0 (very unsatisfied) to 10 (very satisfied)]
		General happiness	Taking all things together, how happy would you say you are? [0 (extremely unhappy) to 10 (extremely happy)]
Psychological wellbeing	Positive characteristics	Vitality	In the past week I had a lot of energy [0 (none) to 3 (all of the time)]
		Emotional stability	(In the past week) I felt calm and peaceful [0 (none) to 3 (all of the time)]
		Optimism	I am always optimistic about my future [0 (strongly disagree) to 4 (strongly agree)]
		Resilience	When things go wrong in my life it generally takes me a long time to get back to normal [0 (strongly disagree) to 4 (strongly agree)] (reverse score)
		Self esteem	In general, I feel very positive about myself [0 (strongly disagree) to 4 (strongly agree)]
	Positive functioning	Engagement	I love learning new things [0 (strongly disagree) to 4 (strongly agree)]
		Meaning	I generally feel that what I do in my life is valuable and worthwhile [0 (strongly disagree) to 4 (strongly agree)]
		Positive relationships	There are people in my life who really care about me [0 (strongly disagree) to 4 (strongly agree)]
		Competence	Most days I feel a sense of accomplishment from what I do [0 (strongly disagree) to 4 (strongly agree)]

by the average of the peak and end of the event (Kahneman, 2011). From a societal perspective, however, it would be difficult to deem a society desirable when the happiness of the young is low and the happiness of the old is high. Thus, different perspectives on the desirability of risk assessment indicators are present for individuals and for society as a whole.

Whether mortality, loss of life expectancy, DALYs, QALYs, happy life expectancy or other indicators should be used ultimately depends on value judgements about what our society strives to achieve. It is not something that is generally set in stone; it is a choice that society must make. One methodology is to introduce an approach of social debate and citizen participation, as discussed later in this chapter. However, it may not be practical in terms of time, opportunity or cost to follow such an approach for all situations. Nevertheless, it is appropriate to use indicators that meet the purpose of the evaluation, are aligned with the worldview citizens want to achieve, are linked to socially meaningful measures and are likely to be shared by many people.

Application of Happy Life Expectancy to Compare Risks between Radiation Exposure and Psychological Distress after the 3.11 Disaster

Thus far, my colleagues and I have examined the following health concerns after the 3.11 disaster: radiation exposure and evacuation from nursing homes, diabetes, psychological distress and effect of returning to one's hometown on wellbeing (Murakami et al., 2018; Murakami et al., 2021; Murakami et al., 2015; Murakami et al., 2017b). In making risk assessments, attention should be paid to how to deal with uncertainty in general. For example, low doses of radiation exposure do not lead to statistically discernible increases in cancer or mortality. However, it is assumed that there is a dose–response relationship, even at low doses, that is similar to that which holds at higher doses. This assumption results in a conservative assessment, meaning that risk estimates may be higher than the actual cancer incidence and mortality (Murakami, 2018). However, uncertainties may arise regarding other risks that are compared to radiation exposure (e.g., evacuation from nursing homes, diabetes). Previous reports have shown that the loss of life expectancy due to evacuation from nursing homes and increased diabetes is greater than that due to radiation exposure, even when risk due to evacuation from nursing homes or increased diabetes was underestimated (Murakami et al., 2015; Murakami et al., 2017b). Regardless of the uncertainty, evacuation from nursing homes in the acute phase and diabetes in the chronic phase are more important issues in terms of loss of life expectancy than the direct effects of radiation exposure. In addition, the cost-effectiveness of diabetes measures has been shown to be better than that of exposure control measures, such as food distribution restrictions and decontamination (Murakami et al., 2017b).

In a prior case study, my colleagues and I conducted a risk comparison between radiation exposure and psychological distress among residents of 13 municipalities subjected to evacuation orders following the disaster, using loss of happy life expectancy as a risk indicator (Murakami et al., 2018). For both risks, only the amount of increase after the disaster was measured. Cancer caused by radiation exposure was calculated using the assumption that it can occur, even at low doses, as described earlier, and the decrease in life expectancy due to cancer incidence was evaluated. However, for psychological distress, only the decrease in emotional happiness was evaluated, although it may also lead to a decrease in life expectancy. Both aspects of a decrease in life expectancy and a decline in emotional happiness could be evaluated on the same scale by using the loss of happy life expectancy indicator. Furthermore, although the onset of cancer itself may cause a decrease in emotional happiness, this part was not included in the study. Sensitivity analysis suggests that if this part were included, the loss of happy life expectancy associated with exposure would be about twice as large.

To assess the risk of psychological distress, future prevalence values are needed. In that study, we assumed that the changes that had occurred up to the time of the study would continue. Two scenarios were assumed: one assuming linear changes (Scenario 1 – S1) and the other assuming exponential changes (Scenario 2 – S2).

As for the target population, the analysis was conducted separately for males and females aged 20, 40 and 65. We found that, regardless of age, gender or the scenario accounting for a change in psychological distress over time, the loss of happy life expectancy due to psychological distress was greater than that due to radiation exposure (Figure 7.1). Among the entire population, the loss of happy life expectancy due to psychological distress was 26 times and 41 times higher, for S1 and S2, respectively, than that due to cancer associated with radiation exposure. The risk of psychological distress was considered greater than that of cancer, even if the possible decrease in emotional happiness caused by cancer development was taken into account (doubled).

These results have implications for what measures should be prioritised in society and what policies are important. First, they suggest the importance of taking into account the risk of psychological distress in addition to radiation exposure in nuclear disasters. The importance of the risk of psychological distress has often been pointed out in the literature on the Chernobyl and 3.11 disasters (WHO 2006; Suzuki et al., 2015), and it is now possible to quantify the magnitude of risk by using the same index. Considering the possibility of future nuclear disasters, global preparedness measures should be developed to mitigate the risk of psychological distress in advance of a disaster.

Second, the results offer insights regarding the priority of measures for Fukushima. Figure 7.1 shows that, under S1 and S2, the risk of psychological distress could be greater in the fourth year after the disaster (which was set as the future estimate) than in the first three years after the disaster. This indicates that there is considerable room to reduce these risks through the implementation of future measures. In other words, the results indicate that measures against psychological distress are important from the perspective of happy life expectancy. It is meaningful to use risk assessment as an approach to change the future.

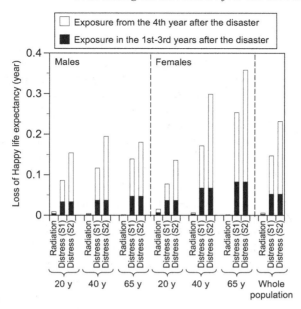

Figure 7.1 Risk comparison between psychological distress and radiation exposure using happy life expectancy (Murakami et al., 2018). S1: linear regression scenario. S2: exponential regression scenario.

How to Determine Which Risk Issues to Prioritise

As we have seen, risk comparisons can provide insights into which risk issues to prioritise. However, priority risk issues are not uniquely determined by the size of the risk or cost-effectiveness of risk reduction alone. Which issues society should solve is a value judgement, requiring us to determine what is important and what kind of world we want to live in (House of Lords Select Committee, 2000). One approach to deciding which issues to solve is to adopt a citizen participation approach. Generally, procedural fairness influences public acceptance (Ohtomo et al., 2020). Therefore, a citizen participation approach often produces a higher level of citizen satisfaction than a top-down decision-making approach.

Here, a citizen participation approach means that citizens are involved in the decision-making process, not that citizens ultimately make the decisions (Murakami et al., 2020b). It is important for experts, the government and citizens to be involved in a step-by-step process, which is determined in advance. Items to consider when developing the process include identifying the stage at which discussions should occur and the extent to which it is acceptable to return to and resume discussions. In addition, to ensure procedural fairness, it is necessary that (1) information is open to the public, (2) there are opportunities for everyone to participate, (3) people who are considered representatives of the citizens can participate, (4) everyone is guaranteed the opportunity to speak, (5) everyone has influence during the final consensus building and (6) the validity of the decision can be evaluated after it is made (Webler, 1995).

These criteria may seem quite obvious, but they are not easy to put into practice. For example, there is often a conflict between items two and three: (2) there are opportunities for everyone to participate and (3) people who are considered representatives of the citizens can participate. This is because when a forum for citizen participation is established with opportunities for everyone to participate, typically those who are very interested in the issue will gather. The participant characteristics are also affected by gender, socioeconomic status, employment status and age, which leads to a bias in terms of representativeness. To solve this dilemma in a practical way, we need to provide two types of forums: one is to provide an opportunity for everyone to participate and the other is to create a forum where people are randomly selected.

Although this approach may take longer than a top-down approach, the level of satisfaction among participating citizens will generally be high (Kweit and Kweit, 2004), which will contribute to progress toward the desired visions of society. Furthermore, it can reduce prolonged conflicts and fragmentation, such as between citizens and the government or between citizens. However, in the aftermath of a disaster, it may not always be possible to implement such an approach when immediate action is required or when resources are not available.

Under such circumstances, one promising approach is to involve those who are working closely with local residents and to extract the visions of society they are aiming to achieve. A previous study extracted the content and purpose of risk communication activities, based on interviews with people who have been practising risk communication activities with residents in Fukushima Prefecture (Honda et al., 2020). According to the results, risk communicators, including medical professionals, a public relations practitioner, a local government official, non-profit organisation workers, a nursery teacher, and a teacher at a *juku* or 'cram' school, established the basic objectives of their activities as 'alleviating anxiety and stress', 'supporting decision-making', 'gaining trust', 'promoting understanding', 'deepening mutual understanding' and 'sharing values and empathy'. The higher level concepts were 'returning to normal life' and 'cultivating a wider perspective'. In particular, 'returning to normal life' consists of elements such as 'residents able to live healthily and happily', 'residents able to feel safe and restoring their confidence', 'bringing back the local community' and 'establishing daily life'. Fulfilment of these elements will be an important requirement for any post-disaster society. In other words, it is important to enable people to live healthily and happily, to live safely and self-reliantly and to re-establish daily life in their restored communities.

Measures that Contribute to Post-Disaster Wellbeing

As discussed earlier, the decline in daily wellbeing that occurs after a disaster is an important issue. Decreased psychological distress and hesitancy to return to one's hometown can lead to a decrease in happy life expectancy that is greater than the decrease in happy life expectancy due to cancer caused by radiation exposure. Identifying the factors associated with the decrease (or increase) in wellbeing is essential for achieving the larger goal of 'returning to normal life'.

Happiness, or wellbeing, is a complex concept that has been subject to robust debate. It is often assumed that wellbeing is a concept that refers to a state of how well someone is, while happiness is an instantaneous feeling, but this is a simplistic summary of a concept that can vary by situation or context. Thus far, I have used the term *wellbeing* without defining it, but here I provide a definition. Wellbeing can be classified into three categories: (1) momentary feelings of emotion, (2) judgements about feelings over the long term and (3) *eudaimonia*, reflecting quality of life in terms of achieving one's highest potential (Nettle, 2005: Graham et al., 2018). The first two are subjective wellbeing, and the third is psychological wellbeing. Subjective wellbeing is based on the subjective judgement of the individual, and this has been used in the field of economics. Momentary feelings of emotions can be measured by the experience sampling method (Csikszentmihalyi and Larson, 1987), a day reconstruction method (Kahneman et al., 2004) or simple questions about emotional experiences from the previous day (Kahneman and Deaton, 2010). Positive momentary emotions (e.g. enjoyment) and negative momentary emotions (e.g. sadness) have different characteristics (Csikszentmihalyi and Larson, 1987; Busseri, 2018) and can be categorised into two subscales.

Judgements about feelings over the long term are assessed by overall satisfaction with life, Cantril's ladder, and other measures (Frey, 2008). These judgements are characterised by an association with income, whereas momentary feelings of emotion do not differ when income exceeds a certain level (Kahneman and Deaton, 2010). By contrast, psychological wellbeing is based on psychological theories and is measured by assessing the components of a good life (Ryff, 2014). One of the representative measurement methods is the European Social Survey, which can be categorised into two scales: (1) 'positive characteristics', as defined by elements such as vitality and resilience, and (2) 'positive functioning', as defined by elements including competence (Huppert and So, 2013). These forms of wellbeing have been used for policy evaluation to improve wellbeing (Kahneman, 2011; Frey, 2008; Huppert and So, 2013).

Below, I present the results of a questionnaire survey of residents in nine municipalities where evacuation orders were lifted after the 3.11 disaster (Murakami et al., 2020a). The previous report categorised (1) momentary feelings of emotion into 'positive emotion' and 'negative-free emotion', (2) judgements about feelings over the long term into 'life satisfaction and general happiness' and (3) psychological wellbeing into 'positive characteristics' and 'positive functioning'. The structured concepts and questionnaire are shown in Table 7.1. Each scale can be scored using categorical confirmatory factor analysis. The most significant findings follow.

First, as previously mentioned, the factor scores for wellbeing were significantly higher for those who had already returned to their hometown than for those who did not know whether they would return. Importantly, this trend was found regardless of the five types of wellbeing. Furthermore, those who decided not to return to their hometown also had significantly higher factor scores for 'life satisfaction and general happiness' than did those who did not know whether they would return, and there was no significant difference between them and those who had returned. This suggests that the state of not knowing whether to return is associated with a decrease in wellbeing. Therefore, from the perspective of wellbeing, it is important to have policies that provide early information about the prospects for daily life so as not to delay the decision-making process regarding returning home.

Second, the disaster-related factors associated with wellbeing differed by evacuation and return status. This indicates that it is important to take measures according to the status of evacuation and return.

Third, psychological distress was mostly not associated with the wellbeing of those who had returned but was significantly associated with those who had been evacuated. This suggests that it is important to take measures to alleviate psychological distress for those who have evacuated.

For those who had returned home, there was a significant association between a history of hypertension and wellbeing. In general, medical services tend to be less accessible in return destinations. Improvement of medical services was considered important in return destinations. In addition, there was a significant negative association between living apart from one's spouse and 'no negative emotions', suggesting the need for social support for those who want to live with their spouse or family but cannot.

For those who wanted to return in the future, factor scores for 'life satisfaction and general happiness' and 'positive characteristics' were lower when there were unemployed people in the household. After the 3.11 disaster, some people lost their jobs due to the evacuation order. For those who returned to their hometown, the results suggest that solving the mismatch between employment supply and demand is important for wellbeing.

Thus, it would be advisable to extract factors related to wellbeing for each evacuation and return situation and to formulate policies according to the characteristics and circumstances of each group or subpopulation.

Conclusion

In this chapter, I introduced the concept of risk and the significance of risk research and then presented examples of various health risks that arose after the 3.11 disaster. Risk is a concept that cannot be separated from the values held by individuals and society and how it is assessed is also related to the setting of goals and the vision of society that is pursued. I then introduced typical risk indicators and their significance. It is important to note that how to assess risk ultimately depends on a social value judgement of what our society aims to achieve. In the meantime, we should use risk indicators that meet the purpose of the assessment, are aligned with the worldview we want to achieve, lead to socially meaningful measures and are likely to be shared by many people. Furthermore, this chapter mentioned the importance of restoring a happy life after a nuclear disaster and introduces an example discussion of measures from the perspective of wellbeing. It is important to note that the factors associated with wellbeing differ for each subpopulation after a disaster, such as evacuation and return status. This suggests the importance of assessing risks and implementing measures tailored to each subpopulation. Simultaneously dealing with numbers about risk and values, as well as evaluating each subpopulation, will allow a post-disaster society to progress toward its goals.

References

Arai, R., Kiguchi, M. and Murakami, M. (2020) A quantitative estimation of the effects of measures to counter climate change on well-being: Focus on non-use of air conditioners as a mitigation measure in Japan. *Sustainability*, Vol. 12 (8694), pp. 1–18.

Bernstein, P.L. (1996) *Against the Gods*. New York: John Wiley & Sons.

Busseri, M.A. (2018) Examining the structure of subjective well-being through meta-analysis of the associations among positive affect, negative affect, and life satisfaction. *Personality and Individual Differences*, Vol. 122, pp. 68–71.

Cohen, B.L. and Lee, I.-S. (1979) A catalog of risks. *Health Physics*, Vol. 36, pp. 707–722.

CNAS - Committee for a New Academic System attached to the Governing Council of the Science Council of Japan (2003) *A new academic system: integration of science and humanities for the benefit of society* [Online]. Available at: http://www.scj.go.jp/ja/info/kohyo/18pdf/1829.pdf. [Accessed: 4 February 2021]. (In Japanese).

Csikszentmihalyi, M. and Larson, R. (1987) Validity and reliability of the experience-sampling method. *Journal of Nervous and Mental Disease*, Vol. 175 (9), pp. 526–536.

Forouzanfar, M.H., Afshin, A., Alexander, L.T., Anderson, H.R., Bhutta, Z.A., Biryukov, S., et al. (2016) Global, regional, and national comparative risk assessment of 79 behavioural, environmental and occupational, and metabolic risks or clusters of risks, 1990–2015: a systematic analysis for the Global Burden of Disease Study 2015. *The Lancet*, Vol 388 (10053), pp. 1659–1724.

Frey, B.S. (2008) *Happiness: A revolution in economics*. Cambridge, MA: Massachusetts Institute of Technology Press.

Gamo, M., Oka, T. and Nakanishi, J. (2003) Ranking the risks of 12 major environmental pollutants that occur in Japan. *Chemosphere*, Vol. 53 (4), pp. 277–284.

GBD 2015 DALYs and HALE collaborators (2016) Global, regional, and national disability-adjusted life-years (DALYs) for 315 diseases and injuries and healthy life expectancy (HALE), 1990–2015: a systematic analysis for the Global Burden of Disease Study 2015. *The Lancet*, Vol. 388 (10053), pp. 1603–1658.

Graham, C., Laffan, K. and Pinto, S. (2018) Well-being in metrics and policy. *Science*, Vol. 362 (6412), pp. 287–288.

Graham, J.D. and Wiener, J.B. (eds.) (1997) *Risk Versus Risk: Tradeoffs in Protecting Health and the Environment*. Cambridge, MA: Harvard University Press.

Graunt, J. (1662) *Natural and Political Observations Made Upon the Bills of Mortality*. London: Royal Society.

Hasegawa, M., Murakami, M., Nomura, S., Takebayashi, Y. and Tsubokura, M. (2019) Worsening health status among evacuees: analysis of medical expenditures after the 2011 Great East Japan Earthquake and nuclear disaster in Fukushima. *Tohoku Journal of Experimental Medicine*, Vol. 248 (2), pp. 115–123.

Honda, K., Igarashi, Y. and Murakami, M. (2020) The structuralization of risk communication work and objectives in the aftermath of the Fukushima nuclear disaster. *International Journal of Disaster Risk Reduction*, Vol. 50, p. 101899.

Hommerich, C. (2012) Trust and subjective well-being after the Great East Japan Earthquake, tsunami and nuclear meltdown: preliminary results. *International Journal of Japanese Sociology*, Vol. 21 (1), pp. 46–64.

Huppert, F.A. and So, T.T. (2013) Flourishing across Europe: application of a new conceptual framework for defining well-being. *Social Indicators Research*, Vol. 110 (3), pp. 837–861.

ICRP (1991) *1990 Recommendations of the International Commission on Radiological Protection. ICRP Publication 60. Ann ICRP 21 (103)* [Online]. Available at: https://www.icrp.org/publication.asp?id=icrp%20publication%2060 [Accessed: 4 February 2021].

International Atomic Energy Agency (IAEA) (2015) *The Fukushima Daiichi accident: report by the Director General*. Vienna: Austria.

ISO (2009) *Guide 73: Risk management — vocabulary* [Online]. Geneva: ISO. Available at: https://www.iso.org/obp/ui/#iso:std:iso:guide:73:ed-1:v1:en [Accessed: 5 August 2021].

ISO/IEC (2014) *Guide 51: Safety aspects - guidelines for their inclusion in standards*. Geneva: ISO.

Johnson, R., Jenkinson, D., Stinton, C., Taylor-Phillips, S., Madan, J., Stewart-Brown, S. and Clarke, A. (2016) Where's WALY?: a proof of concept study of the 'wellbeing adjusted life year' using secondary analysis of cross-sectional survey data. *Health and Quality of Life Outcomes*, Vol. 14 (1), pp. 126–126.

Kagan, S. (2012) *Death*. New Haven: Yale University Press.

Kahneman, D. (2011) *Thinking, Fast and Slow*. New York: Farrar, Straus and Giroux.

Kahneman, D. and Deaton, A. (2010) High income improves evaluation of life but not emotional well-being. *PNAS*, Vol. 107 (38), pp. 16489–16493.

Kahneman, D., Krueger, A.B., Schkade, D.A., Schwarz, N. and Stone, A.A. (2004) A survey method for characterizing daily life experience: the day reconstruction method. *Science*, Vol. 306 (5702), pp. 1776–1780.

Kanoshima, E. (2000). The concept of risk and the problems of present society. *Japanese Journal of Health Physics*, Vol. 35, pp. 473–481.

Kweit, M.G. and Kweit, R.W. (2004) Citizen participation and citizen evaluation in disaster recovery. *American Review of Public Administration*, Vol. 34 (4), pp. 354–373.

Maeda, M. and Oe, M. (2017) Mental health consequences and social issues after the Fukushima disaster. *Asia Pacific Journal of Public Health*, Vol. 29 (2S), pp. 36S–46S.

Murakami, M., Ono, K., Tsubokura, M., Nomura, S., Oikawa, T., Oka, T., et al. (2015) Was the risk from nursing-home evacuation after the Fukushima accident higher than the radiation risk? *PLoS one*, Vol 10 (9), p. e0137906.

Murakami, M., Harada, S. and Oki, T. (2017a) Decontamination reduces radiation anxiety and improves subjective well-being after the Fukushima accident. *Tohoku Journal of Experimental Medicine*, Vol. 241 (2), pp. 103–116.

Murakami, M., Tsubokura, M., Ono, K., Nomura, S., Oikawa, T. (2017b) Additional risk of diabetes exceeds the increased risk of cancer caused by radiation exposure after the Fukushima disaster. *PLoS one*, Vol. 12 (9), p. e0185259.

Murakami, M. (2018) Importance of risk comparison for individual and societal decision-making after the Fukushima disaster. *Journal of Radiation Research*, Vol. 59 (Supplement_2), pp. ii23–ii30.

Murakami, M., Tsubokura, M., Ono, K. and Maeda, M. (2018) New "loss of happy life expectancy" indicator and its use in risk comparison after Fukushima disaster. *The Science of the Total Environment*, Vol. 615 (2018-02-15), pp. 1527–1534.

Murakami, M. (2019) *Risk and value - from experiences in Fukushima: SYNODOS* [Online]. Available at: https://synodos.jp/fukushima_report/22635 [Accessed: 4 February 2021]. (In Japanese).

Murakami, M., Takebayashi, Y., Ono, K., Kubota, A. and Tsubokura, M. (2020a) The decision to return home and wellbeing after the Fukushima disaster. *International Journal of Disaster Risk Reduction*, Vol. 47 (2020-08), p. 101538.

Murakami, M., Ohnuma, S., Hayashi, T.I., Tsugane, S. and Yasutaka, T. (2020b) Deepening and updating of risk analysis from values and norms. *Japanese Journal of Risk Analysis*, Vol. 29 (3), pp. 205–209. (In Japanese).

Murakami M, Takebayashi Y, Ono K, Tsubokura M (2021) Risk trade-off analysis of returning home and radiation exposure after a nuclear disaster using a happy life expectancy indicator. *Journal of Radiation Research*, Vol. 62 (Supplement_1), pp. i101–i106.

Murray, C.J., Lopez, A.D. and Jamison, D.T. (1994) The global burden of disease in 1990: summary results, sensitivity analysis and future directions. *Bulletin of the World Health Organization*, Vol 72 (3), pp. 495–509.

Nettle, D. (2005) *Happiness: The science behind your smile.* Oxford: Oxford University Press.

Nomura, S., Blangiardo, M., Tsubokura, M., Ozaki, A., Morita, T. and Hodgson, S. (2016) Postnuclear disaster evacuation and chronic health in adults in Fukushima, Japan: a long-term retrospective analysis. *BMJ open*, Vol. 6 (2), p. e010080.

Nomura, S., Gilmour, S., Tsubokura, M., Yoneoka, D., Sugimoto, A., Oikawa, T., Kami, M., Shibuya, K.H. and Chuhsing, K. (2013) Mortality risk amongst nursing home residents evacuated after the Fukushima nuclear accident: a retrospective cohort study. *PLoS one*, Vol. 8 (3), p. pe60192.

Ohtomo, S,. Hirose, Y. and Ohnuma, S. (2020) Public acceptance model for siting a repository of radioactive contaminated waste. *Journal of Risk Research*, Vol. 24 (2), pp. 215–227.

Rehdanz, K., Welsch, H., Narita, D. and Okubo, T. (2015) Well-being effects of a major natural disaster: the case of Fukushima. *Journal of Economic Behavior and Organization*, Vol. 116 (2015–08), pp. 500–517.

Ryff, C.D. (2014) Psychological well-being revisited: advances in the science and practice of eudaimonia. *Psychother Psychosom*, Vol. 83 (1), pp. 10–28.

Sawano, T., Nishikawa, Y., Ozaki, A., Leppold, C. and Tsubokura, M. (2018) The Fukushima Daiichi Nuclear Power Plant accident and school bullying of affected children and adolescents: the need for continuous radiation education. *Journal of Radiation Research*, Vol. 59 (3), pp. 381–384.

Society for Risk Analysis Japan [ed.] (2019) *The encyclopaedia of risk research*. Tokyo: Maruzen. (In Japanese).

Sone, T., Nakaya, N., Sugawara, Y., Tomata, Y., Watanabe, T. and Tsuji, I. (2016) Longitudinal association between time-varying social isolation and psychological distress after the Great East Japan Earthquake. *Social Science & Medicine*, Vol. 152 (2016-03), pp. 96–101.

Suzuki, Y., Yabe, H., Yasumura, S., Ohira, T., Niwa, S., Ohtsuru, A., et al. (2015) Psychological distress and the perception of radiation risks: the Fukushima Health Management Survey. *Bulletin of the World Health Organization*, Vol. 93 (9), pp. 598–605.

Tiefenbach, T. and Kohlbacher, F. (2015) Happiness in Japan in times of upheaval: empirical evidence from the national survey on lifestyle preferences. *Journal of Happiness Studies*, Vol 16 (2), pp. 333–366.

Tsuchiya, A., Dolan, P. and Shaw, R. (2003) Measuring people's preferences regarding ageism in health: some methodological issues and some fresh evidence. *Social Science & Medicine*, Vol. 57 (4), pp. 687–696.

House of Lords Select Committee on Science and Technology (2000) *Science and technology - third report* [Online]. Available at: https://publications.parliament.uk/pa/ld199900/ldselect/ldsctech/38/3801.htm [Accessed: 4 February 2021].

United Nations Scientific Committee on the Effects of Atomic Radiation (UNSCEAR) (2014) *UNSCEAR 2013 report vol. 1: sources, effects and risks of ionizing radiation.* New York: United Nations.

Webler, T. (1995) "Right" discourse in citizen participation: an evaluative yardstick. In Renn, O., Webler, T. and Wiedemann, P. (eds.) *Fairness and competence in citizen participation: evaluating models for environmental discourse.* Dordrecht: Kluwer Academic Publishers, pp. 35–86.

Williams, A. (1997) Intergenerational equity: An exploration of the 'fair innings' argument. *Health Economics*, Vol. 6 (2), pp. 117–132.

World Health Organization (WHO) (2006) *Health effects of the Chernobyl accident and special health care programmes: Report of the UN Chernobyl Forum Expert Group "Health"* [Online]. Available at: https://www.who.int/publications/i/item/9241594179 [Accessed: 4 February 2021].

Zeckhauser, R. and Shepard, D. (1976) Where now for saving lives? *Law and Contemporary Problems*, Vol. 40 (4), pp. 5–45.

8 Commensurability and Post-Disaster Mental Health after 3.11

Ben Epstein

Introduction

This chapter is based on ethnographic fieldwork with mental health practitioners, researchers, and stakeholders after the Tōhoku, March 2011 earthquake, tsunami, and nuclear disaster. The chapter explores conceptual and epistemological questions about knowledge and practice in post-disaster mental health (DMH) using the concept of commensurability. This reveals the ways that different entities are deemed comparable in post-DMH. The practice of psychological humanitarian aid highlights how experiences in one place are made commensurate with experiences in another through the production and reproduction of legitimate scientific authority at each disaster location. The need to produce universal, standardised assessments of health and wellbeing is driven by the need for organisations to obtain and secure funding, by international criteria, and the development of novel technologies. I bring attention to the effects of commensuration, particularly how data are collected and used to reify the experience of survivors.

I examine two cases in this chapter. The first characterises the increasing power of biomedical interventions into mental disorders, reflected in government financial support and the promise of creating instrumentalised biotechnologies beyond the immediate situation. The second case regards a type of mental health and psychosocial support (MHPSS) service made available in Japan after the 1990s' Kobe earthquake and diffused after the recent disaster, locally known as *Kokoro no Kea*. This forms the bulk of the focus in the latter part of the chapter. I then inform my analysis with a theoretical perspective on how and why the hinterland of practitioner resources may not be captured by numerical data, drawing on the approach of psychosocial supporters in *Kokoro no Kea* (care of the heart/mind) to respond adaptively to the needs of survivors on the ground.

The role of mental health services in disaster and war-affected areas, particularly those which focus on building resilience, has been criticised in various contexts as broadly being depoliticising, dehumanising, and pathologising of people's needs and experiences. An extension of what Vanessa Pupavac has called 'therapeutic governance', referring to 'the management of a population's psychology, and its significance for security' (Pupavac 2005:1), creating

DOI: 10.4324/9781003182665-10

resilient communities also means creating communities insulated from within to the increasing demands of flexible forms of capitalism and resilience in the face of declining social and welfare services. Mass-produced psychotherapy, such as cognitive behavioural therapy (CBT) or mindfulness-based CBT (which can be administered online or without input from specialists) can have the unintended consequence of putting the responsibility 'to change' on the patient or client rather than on social injustices (see also Tierney 2015). In Fukushima, this can translate into the pathologisation of people's (legitimate) fear of radiation.

After the 3.11 disasters, the private and voluntary sectors hastened to respond to the emerging needs of survivors in areas with relatively limited access to mental health provisions. Correspondingly, the effects of the disasters have bolstered calls for policy reform of the regional (and national) mental health system in ways that reflect impulses towards creating more sustainable (and cost-effective) community-based services (cf. Richard 2012) such as psychosocial support. Beyond a general sense that survivors will inevitably suffer from mental health issues, data-driven epidemiological studies have propelled risk-focused needs assessment into the prediction and management of risk. As 'the logic of prediction comes to replace the logic of diagnosis' (Rose 1998:185), fewer mental health professionals are involved with actively diagnosing psychiatric illness, and the focus is now much more on symptom risk management and nebulous 'support' for mental health and psychosocial wellbeing. Given that the volume of activity in the aftermath of a disaster is overwhelming it is unsurprising that researchers, policymakers, and clinicians reach for established 'off the shelf' tools and strategies that will make the process of classifying people into manageable categories for which resources can be eked out easier. It became clear to me in my fieldwork that an underlying assumption by those trying to help was that if a tool worked for one place, time, event, or age group, it surely must work for another. Global institutions and the individuals who put the most relevant and up-to-date diagnostic criteria to use make previously unknown disorders 'visible and quantifiable in a way that is *commensurate with global scientific discourse*' [emphasis added] (Breslau 2004:117). This is what is taking place when individual stories are aggregated and re-aggregated to fit standardised models of psychiatric disorder and psychological distress.

For the person who witnesses a disaster, the experience is real and lived, but are extant psychological tools really able to grasp such experiences? An ethnographically grounded theory of commensurability could generate useful and robust understandings about what underpins data-driven policy, research, and practice in DMH care. Are mental health and psychosocial actors aware of the inherent flaws in any system that attempts to quantify experience, particularly in the screening phase before interventions have been applied?

Generalised forms of MHPSS include providing advice on mental health issues and providing individuals with day-to-day concerns such as childcare and employment. Alongside these, potential treatments used amongst disaster-affected populations in Japan include trauma-focused cognitive-behavioural therapies (CBTs), eye movement desensitisation and reprocessing (EMDR), and, to a lesser extent, pharmacological treatments, specifically selective serotonin reuptake inhibitors (SSRIs). In parallel to traditional MHPSS activities laboratories around Japan are conducting basic and clinical research to investigate the molecular mechanisms of disaster stress responses and disaster-related mental disorders (Bowers and Yehuda 2016; Takahashi et al., 2016). This is specifically aimed at quantifying markers of trauma after exposure to putatively traumatogenic events using biomedical methodologies (Müller et al., 2017). To date, however, little work has taken this body of scientific endeavour back to first principles. Notably, doing so would reveal that psychological instruments used to identify target populations are tenuously standardised and may not have been thoroughly validated within the local disaster context (Tol et al., 2011). This produces a chasm between mental health research and practice.

The biology of post-traumatic morbidity is emphasised in DMH research, in line with developments in biomedical research globally (Friedman et al., 2007; Raphael and Maguire 2009). Despite refinements in the reliability of diagnostic criteria, there have been no parallel improvements in the validity of diagnosis (Moncrieff and Timimi 2013). To improve validity, some researchers are hoping to transform the way post-traumatic stress is diagnosed in the future by assessing mechanisms in the brain and underlying biological markers, and not simply from a symptom checklist. The hope for mental health 'lies in extending this research to meaningful strategies that might contribute to neuroprotection, neuroplasticity, and resilience and that may make for pathways to prevention of PTSD [post-traumatic stress disorder] and other post-trauma morbidity' (ibid).

Despite such advancements, it appears there is a disconnect between DMH research and governance (with its voluminous auditable data possibilities) and DMH in practice. It is this disconnect that this chapter engages with, through the case study of mental health research and support following 3.11. Assessing the effectiveness of MHPSS intervention outcomes after disasters anywhere is notoriously complex, let alone after the 3.11. Contrary to what I was told by a molecular psychiatrist who I worked with that '[t]here is no difference between research and practice in disaster mental health', there is a chasm between the roles of researchers and practitioners. At the disaster psychiatry research institute where I worked, events were held to be a priori traumatising (cf. Fassin and Rechtman 2009) and the embodied experiences of those classed as survivors were extracted from the body and then entered the laboratory as data and harvested tissue. But, as Ecks puts it, 'modern psychiatry has failed to integrate symptom classifications with biomarkers' (Ecks 2018:01) which means that innovation in classification has not been able to keep up with the development of biotechnologies designed to locate the biological markers of mental disorders in the body. He describes this as 'a failure of biocommensurations' (Ecks 2018:01). Indeed, there are highly complex assemblages within diagnostic categories used to delimit mental health from disorder, and these rely on the commensurability of personal information:

accumulated stores of harvested data that can be analysed in terms of their internal systemic properties unmoored from the quotidian contexts people inhabit and largely inattentive to political, social, or environmental flux. In post-DMH research, 'social contexts' (factors difficult to pin down or operationalise) are made a general feature of clinical practice, turned into study sets, fungible data, and medical experience. Commensurability refers to the ways in which the relevance of context can be reduced (or deleted) through standardisation: 'qualities become quantities, differences magnitude' (Espeland and Stevens 1998). I argue that commensuration forms the epistemological groundwork for DMH. By groundwork, I mean the taken-for-granted background of decision-making that requires considerable efforts to maintain (Law 2004).

Within this context, samples are collected from individuals screened for post-traumatic reactions and reformulated in the lab by clinicians and researchers in precision psychiatry (cf. Fernandes et al., 2017; Gómez-Carrillo et al., 2018). Then the genetic component of traumatic stress has been identified, captured, and mapped onto or cross-validated with psychometric instruments and other measurement strategies. A key question that needs to be asked is this: Beyond euphemistic appeals to more 'personalised' treatment, what kind of care will be created?

The Fieldwork

In this section, I briefly provide a background outline of the project I undertook. In all the accounts given, individuals and their workplaces have been anonymised to protect their identities. Where this is not the case, pseudonyms are used. Anonymisation includes changes to identifiers such as occupation, family background, gender, and even amalgamating statements from multiple participants to create a singular identity. I 'followed' my people (mental health workers), the 'things' (mental health), and the 'assemblage' (of actors and knowledge practices), as it were, across a number of different settings, most of which were institutional or organisational. Although this fieldwork was multi-sited, I connected these places less in terms of their intrinsic differences and more as a pattern of the broader thematic category of DMH as 'a point of potentiality and multiple possibility' (Kapferer 2005:89). This could access the seemingly disparate ways of dealing with the mental health of those affected, that is, through telephone technologies, local clinics, outreach, non-profit volunteer organisations, biomedical research laboratories, and other forms of 'data' or knowledge work. What united these sites and activities was not so much the similarity of their approaches but their commitment to aspects of 'post-disaster mental health'.

After reading about a possibly unfolding 'mental health crisis' after the earthquake (Fukunaga and Kumakawa 2015), I decided to explore ways in which ethnographic fieldwork might be used to generate insight to support mental health in the region. I was keen to go out into the 'field' – 'that dubious category that has come to signify any shift in location that is the ethnographic pretext' (Mosse

2006:936). To do so, I became a visiting scholar at a cutting-edge research department in Tōhoku specialising in disaster psychiatry with a view to gaining proper authority to access local clinicians for my fieldwork. Although I had not set out with the intention to study the workings of the department itself (henceforth, the 'lab'), when I arrived there my study took a surprising turn. My host requested that I conduct a small-scale pilot study of the lab and of the lab workers before being allowed to enter the 'field' I had initially intended to explore. Thus, the lab became a key part of my project. This would create a situation of conflicting interests on several levels, demonstrating what is 'ethnographically valuable' to a supervising clinician, researching a group of clinicians and laboratory workers with whom I was closely affiliated, and grappling with the fact that many of the researchers were not actively engaged in clinical work with people from the disaster area. Work in the lab appeared to be organised in a fairly loose and individualised manner. It turned out that outsiders were expected to follow protocols put in place for clinical researchers, and I was restricted in regards to whom or when I could interview staff members.

This fieldwork experience reflected some of the wider concerns around DMH. In a sense, the operation of the lab was a microcosm of the broader challenges: between disaster risk policymakers, mental health researchers, and clinicians and the conceptual framework each may be using to interpret the reality of victims into interventions. Different actors transform data into activity based on a complex interplay of commensurations, negotiating, and prioritising often competing factors. For example, in a study among tsunami-evacuated residents aiming to inform local policy on deciding where to relocate a population based on mental health outcomes, who should be included or excluded? The study highlighted that regardless of whether people go back to the area they came from or move to new areas, there is no difference in levels of depressive symptoms or PTSD. Despite any issues there may be about the research design of this specific example, it was indicative of a broader pattern in DMH research in which 'averages' can start to take primary focus, without all diverse members of a community necessarily being recognised, that is, 18-year-olds and 80-year-olds can become an average 57.0-year-old person. It also highlights how symptoms and experiences become categorised. Post-traumatic reactions and depressive symptoms experienced by individuals can be quantified to become a single number.

A form of pragmatism was evident in the lab where delivering deadlines on time took priority over data sharing and joining up work. Overall, whether in the lab with researchers or in the community with clinicians, it seemed that despite the best intentions of those attempting to bridge the gap between explicit and implicit knowledge, practice, and data, results are still unpredictable and is difficult to effect policy change.

My presence at the lab was 'pre-approved', and the purpose of my research had been accepted. Initial exploratory work was planned to gain insight into local DMH practices using transcribed audio recordings from ethnographic interviews and field notes for textual analysis. It seemed that there was an implicit agreement that we were working towards a common goal. So it was surprising that in practice, it would take more than a year before I was able to complete a second independent review board(IRB) clinical research ethics protocol to be able to interview the members of the laboratory in order to obtain permission to conduct fieldwork with mental health practitioners outside the lab with whom I had previously negotiated access.

Perhaps my hosts were suspicious of my intentions as an ethnographer with an open-ended research programme. What might I uncover that they were not prepared for? Would I have the same values or operate from the same tacit frameworks that they did? Through the ethical review process, they took steps to make explicit my intentions via integrating surveys and structured interview questions into my research.

In a circuitous way, this experience led me to narrow down my research focus on the concept of commensurability. As I struggled to understand where the anxiety about my work was coming from, it became evident that both the open-ended way I was approaching my research and the content of my questions may inadvertently reveal information that current data collection tools in use at the lab were either missing entirely or unable to capture. For example, my research project was intended not to be predetermined at the outset. I could not decide on a questionnaire until I had done preliminary observations and unstructured interviews and allowed themes to emerge, whereas the typical approach of the lab was to be goal-focused. This involves having a system in which questions should be decided beforehand so that any answers could be systematised and encoded using spreadsheets and software, that is making the data produced *commensurable*. This was perhaps partly a failure of transitivity on my part: to explain my intentions in terms amenable to the clinical researchers I worked with.

My approach, in being adaptable, did reveal issues with the system. For example, when I later talked to participants, there was broad agreement in the intention to 'do good' but no consistency or clarity about which actions would have the most impact, long-term or short-term, across multiple discriminating factors.

Other tensions arose around researching in a concrete and calculable way rather than in an open-ended ambiguous way. The lab sought to control my movements, my interviews, and my findings. Finally, ethnographic arguments often are under a 'double burden' of scrutiny as these 'tend to be assigned an inferior position in the hierarchy of evidence that increasingly shapes decision-making in the health services' (Savage 2006:390). The fact that my role as an ethnographer was unlike other roles at the organisation unavoidably set me apart, either as a benign 'unknown quantity' or potentially a time waster doing activities not recognisable as scientifically or clinically valuable.

Commensurability in Practice

Commensurability speaks to the *capacity* of something to be made comparable to something else or to institute, or be the source of, comparison. Commensuration, on the other hand, raises the issue of what processes are at play when practitioners hold or bracket the ontological basis of distress 'in suspense' precisely because of the narrative complexity of illness experience and clinical care that often eludes concretisation in universalistic disease categorisation. As Espeland and Stevens put it,

> [c]ommensuration is the expression or measurement of characteristics normally represented by different units according to a common metric. Most quantification can be understood as commensuration because quantification creates relations between different entities through a common metric. Everyday experience, practical reasoning, and empathetic identification become increasingly irrelevant bases for judgment as context is stripped away and relationships become more abstractly represented by numbers.
>
> (Espeland and Stevens 1998:315)

Before turning to the question of DMH practice, it is worth spending a few words here on the concepts of commensuration and commensurability because these terms will shape the focus of the discussion. *Commensuration* in this case can be broadly explained as the process of comparing entities or properties using frameworks and tools to abstract into data. For example, experiences and symptoms identified within diagnostic frameworks in one disaster are selected and evaluated as commensurable to those collected in another, making two different disaster events comparable on a scale. Commensurating is often taken for granted as a necessary means to enable authorities to prioritise resources in difficult situations.

The umbrella term *mental health and psychosocial support (MHPSS)* is used by stakeholders to describe all manner of interventions after disasters. MHPSS interventions exist on a continuum, with more professionalised and accountable services provided by mental health services at one end and less objectively technical psychosocial support interventions or activities on the other. These have become 'a standard component of humanitarian response' in disaster settings around the world (Tol and Van Ommeren 2012:25–26) raising legitimate questions regarding the validity and specificity of transplanting psychiatric and psychological nosology and diagnosis into novel contexts – often with minimal local consultation. All this activity (funded by international, national, or voluntary organisations) inevitably leads to a plethora of interventions and data collection strategies, which need to be quantifiable to justify the spending, if nothing else.

Ostensibly operating within an evidence-based framework, a number of mental health interventions have been rolled out and field-tested in Japan during this crisis. Initially used with adults and children exposed to disaster trauma in other communities around the world (Chernobyl is often cited), these included cognitive processing therapy, interpersonal psychotherapy, prolonged exposure, narrative exposure therapy, and trauma-focused cognitive behavioural therapy (Shultz et al., 2011). A crucial factor in determining who is given access to such services, and thereby which therapies, is determined in the initial screening and the preferences and abilities of an individual practitioner or organisation.

Screening Scales and Diagnostic Tools

Psychological assessment scales and measures draw on contingent assumptions about the reliability and validity of psychological data in disaster contexts. Such instruments arguably pay insufficient attention to the personal resources care workers frequently draw on to co-produce beneficial outcomes as determined by the person in need of support and which are omitted in statistical models. Such models then may go on to attribute improved outcomes to screening and diagnostic interventions instead.

Measurements, indexes, and probabilistic methods used in psychological testing are seen as an effective way to quickly generate data relevant to public policy and health administration – particularly in resource-limited emergency situations – but are deterministic to the extent that they produce substantive claims about the factors at play.

An example of where I observed screening tools used in practice was at a call centre. Focused on providing psychosocial support, daily work for the consultant team involved spending many hours on the telephone going through a list of phone numbers of residents who live in the disaster-affected communities. The individuals selected to call are amongst those considered most at risk as calculated from the initial screening and psychological survey distributed in a health questionnaire. Although I was understandably not able to obtain copies of the completed questionnaires used, the most widely accepted (and the only 'validated') questionnaire is the Kessler Psychological Distress Scale (K6/K10). The K6 is an abbreviated version of the K10, a widely used measure for either screening or severity of mental distress.

The following is an excerpt from an interview with a telephone counsellor I will call Dr Umeda, who was asked about the work she was performing:

> Dr Umeda: When making calls, firstly, I check people's scores. People with high scores will have been expecting a call for a while and they might be in a serious condition. When I make a phone call, at first, everyone seems to answer the phone in a lively voice. So, I introduce myself and explain for what reason I am calling. There was a time when someone's depression score was really high. After I called, they sounded worried. When I started this work, I used to tell people I was with the clinic, and they would start getting really alarmed. I don't think these calls help people speak honestly or directly. Now, I do not even mention mental health. I just say that I am a clinical psychologist working to promote health. I try not to use technical terms as much as possible like depression or mental health. I say I'm just calling to talk about their heart.
>
> *(kokoro)*

Using the Kessler 6 Screening Scale for Psychological Distress to measure affective or anxiety disorder, the Fukushima Mental Health and Lifestyle Survey, part of the Health Management Survey initiated by Fukushima University, is 'the only study to use a verifiable indicator of mental wellbeing of registered residents in Fukushima Prefecture' (Fukushima, 23/04/19). This scale, which ranges from 0 to 24, asks respondents whether they have experienced six specific mental health symptoms.

Respondents rate how often they felt nervous, hopeless, restless, or fidgety, so sad that nothing could cheer them up, that everything was an effort, and feelings of worthlessness during the past 30 days. In assessing the utility of the K6, two types of studies are common. First, the ability of the Japanese translation of the K6 to predict *Diagnostic and Statistical Manual of Mental Disorders (DSM)*–based diagnoses for mood and anxiety disorders has been explored in a wide range of community samples. It has been deployed many thousands of times. Second, investigators have examined the degree to which the K6 is related to known correlates of severity. 'The outcomes of these efforts have indicated that the K6 is both an effective screening measure and an indicator of distress severity among the populations that have participated to date' (Mitchell and Beals 2011:754).

A modified version of such an instrument was thus deployed in Fukushima evacuees (Suzuki et al., 2015a). Participants' beliefs about the potential health effects of radiation exposure were based on their responses to the following questions: (i) What do you think is the likelihood of having immediate health damage (e.g., dying within one month) as a result of your current level of radiation exposure? (ii) What do you think is the likelihood of damage to your health (e.g., cancer onset) in later life as a result of your current level of radiation exposure? (iii) What do you think is the likelihood that the health of your future (i.e., as yet unborn) children and grandchildren will be affected as a result of your current level of radiation exposure? 'These items were translated into Japanese, then back to English, and modified after discussion with the authors of the questionnaire' (Suzuki et al., 2015b:600). Participants were asked to respond to each question using a four-point Likert scale as follows: very unlikely (1), unlikely (2), likely (3), or very likely (4).

After this, the data from the survey are modelled using a statistics package such as SPSS. Demographic data are superimposed using frequencies and individuals with scores lower than or equal to 5, or in some cases 13 or 10, are cut off. An individual with higher K6 scores of over or equal to 5 (again, in some cases) was considered as having experienced psychological distress.

So, what can we learn from these results? The unpredictability of this type of instrument seems self-evident, not least the decision of when and where to set the cut-off point. Medical professionals must negotiate the tension between distancing quantitative technologies – applying structured diagnostic instruments, surveys, and screening tools – and the primacy of embodied experience (Csordas and Harwood 1994). There is a productive tension between the need for the medical professional to represent experience as data and the lived experience itself, which cannot entirely be constructed via representation. This reveals how science and technology construct analytical categories that are at once black-boxed and florid. Such methods could be said to 'enact' realities as well as describe them (Law 2009).

Again, this raises issues of commensurability, as such surveys used and applied to affected populations rely on statistical inferences that create a homogeneous group, *qua* traumatised individual, based on the assumption that 'traumatic events' are themselves equivalent (allowing for the intensity, duration, and type of disaster agent). Second, this more or less homogeneous population, which exists primarily in the literature, can be 're-stratified' into sub-spaces or sub-populations to allow for epidemiological differences across a set of predetermined ahistorical factors such as age, gender, class, or even levels of 'social support' (Suzuki et al., 2015a). Ultimately, the numbers are analytical similes by which clinicians may test their assumptions (Danziger 1994). It should be noted that some psychiatrists are well aware of the limitations of the K6 used both to measure 'non-specific' psychological distress and as a screening tool. For example, Suzuki et al. (2015b:603) state that the results 'cannot be generalized to evacuees in other disasters or other populations under normal circumstances, as each disaster has different features, and the affected community has different social and demographic characteristics'.

Many DMH researchers acknowledge the need to control for the cumulative effects of disasters on affected populations. After the Chilean Bío Bío Earthquake, researchers (Garfin et al., 2014) used multivariate analyses to untangle demographic effects from distress and post-traumatic stress responses to disaster exposure and secondary stressors. In doing so, statements about which kinds of disasters may be deemed 'worse', in contrast to those that were most distressing, can be made. This allows researchers to account for different types of events by severity and number of events experienced. However, the culturally laden components of what counts as 'community', when comparing the effects of one disaster and another, is rarely sufficiently addressed. Whilst not all quantitative researchers see traumatic events as strictly equivalent, the way data is extracted from one disaster to support investigations into another disaster implies that there is an underlying comparability, either as similarity or difference between these events and others. However, it is worth considering what effects such comparisons have on populations themselves, to ask what, if anything, is lost or gained in the process of commensuration.

In the case of 3.11, the Fukushima Health Management Survey, which studied long-term low-dose radiation exposure caused by the accident, includes an assessment of mental health and lifestyles of all residents from the evacuation zones and records of all pregnancies and births among all women in the prefecture who were pregnant on 11 March (180,604 people received the questionnaire; Suzuki et al., 2015b). With almost every aspect of health and everyday activities being closely monitored, there is a sense, especially in Fukushima Prefecture, of people experiencing what one professional described to me as 'survey fatigue'.

As the disaster has faded from popular attention, the number of volunteers coming to northeast Japan has plummeted, and many not-for-profit organisations face bankruptcy and the shutdown of their activities for want of donations and staff. Professor Masaru Mimura, Dr Hiroyuki Uchida, and Dr Yutaka Kato, psychiatrists at the Department of Neuropsychiatry at Keio University, responsible for providing medical support to Soma City,

Fukushima Prefecture, in 2011, wrote of being unable to accurately monitor the changes in the mental state of the victims and to provide appropriate cures and care for their conditions even in the immediate aftermath of the disasters when more funding was presumably available:

> In fact, we found that the situation was beyond the capacity of the medical university staff and volunteers ... medical support teams could not give long-term help due to the limitation of the schedule and staff fatigue.
>
> (Kato et al., 2012:17–21)

Screening scales and diagnostic tools are intended to group issues together, making data into smaller and more manageable chunks. It would, of course, be an unattainable goal to try to process all the data that make up 'a life span in all its temporal and spatial expanse' (Stolz 2018); instead, the goal seems to be one of parsimony: How can we produce predictable outcomes using the least amount of data and resources possible? One practitioner who ran a clinic in Fukushima said, 'There is a difference in how to support remote areas isolated from the surroundings and how to support in urban areas, and the way of support varies depending on the time elapsed after disaster. Because the mental health and health needs of affected areas are influenced by various factors, a uniform model is not practical. From this point of view, it seems practical to share basic principles'.

There is something to be said for the promise of psychometric instruments and psychiatric inventories to facilitate health and care under conditions of scarcity. But after these tools and interventions disappear from lack of funding or evidence of efficacy the communities obtain support where they can in disparate ways. Where do people go for help when the more formal structures break down?

Kokoro no Kea

The perceived need to 'saturate the area with psychiatrists' at the time of the Kobe earthquake led to the development of specialised clinics labelled *Kokoro no Kea Senta-* (literally, 'care for the heart' centres) to recover the spirits of 'ordinary people suffering from diverse subclinical symptoms' and traumatic reactions (Breslau 2000:179). The first generation of these organisations founded in 1995 'was charged with studying and teaching about *Kokoro no Kea* and providing services to earthquake victims' (Breslau 2000:181). As Breslau points, *kokoro* emerged as a powerful metaphor in contrast to the more widely diffused *seishin* (translatable as 'psyche', as in *seishin igaku*, psychiatry). Takeo Doi's translation of *kokoro* includes concepts of intention, emotion, thought, mind, heart, and subjectivity (Doi 1988: 27, 44, 107 cited in Breslau 2000:181):

> Whereas *seishin* is a word formed with Chinese characters, carrying an academic tone, *kokoro* is derived from an ancient Japanese word, found

in the earliest collections of Japanese writings. While *kokoro* is not a term generally used to refer to oneself, neither is it a technical term. It is an abstract term for comprehending individual motivations and actions, a marked 'Japanese' approach to what is universally human. Paired with the English-derived care, kokoro gives an ancient pedigree to an innovative practice.

(Breslau 2000:182)

Kokoro no Kea has since come to encompass a range of interventions, including psychological first aid (PFA), art therapy, group counselling, and ostensibly 'non-medical interventions' including, but not limited to, facilitating meetings and recreation, outreach services, and day-care for children. *Kokoro no Kea* offers an alternative vision of DMH than the one promoted by large-scale research into post-disaster psychological morbidity. It is instead concerned with 'dealing with psychological ideas of suffering of a generalised nature' (Breslau 2000:181–182) and offering a less stigmatising model of mental health than that suggested by *seishin*. Despite the abundance of therapies on offer throughout the many *Kokoro no Kea* Centres I visited, there is still a belief that 'widespread stigma toward mental illness has complicated the "recovery" process' (Karz et al., 2014:113). Highlighting 'the gap between statistical risk figures and phenomenological experience' (Rapp 1999:175) – despite statistical evidence of a considerable rise in mental illness in the region – survivors are said to be reluctant to seek mental health support 'likely preventing clinically distressed people suffering from seeking help' (Karz et al., 2014:113) Stigma against people with mental disorders is still high among the public (Kim 2018; Setoya 2012:10), and *Kokoro no Kea* is often seen as a more palatable alternative to the English term *mental health* (Harding 2017).

The origin of centres like the ones I visited in Fukushima began in the immediate aftermath of the Dai-Ichi Nuclear Power Plant incident. Several hospitals with psychiatric inpatient facilities within a 30-km radius of the nuclear power plant were closed and ordered to move their inpatients to other hospitals outside the 30-km zone. As a result, more than 800 inpatients were transferred to other hospitals within or outside Fukushima Prefecture. In addition, three psychiatric clinics within the radius stopped operating entirely, leaving a large gap in service provision (Niwa 2014:621).

Kokoro no Kea centres were either opened by universities or set up as not-for-profit organisations funded by the governments' reconstruction budgets (Seto et al., 2019). They have the advantage of being 'drop-in' as well as providing outreach, allowing for a model of care that can respond to changing needs day-to-day and even year to year. However, they are seen only as a temporary solution. One can speculate for a number of reasons, such as the challenge for *Kokoro no Kea* centres to provide measurable, quick results over short periods. Interventions are not standardised in the same way as hospital care and therefore are perhaps harder to fund. In an overview article on the MHPSS activities after the Great East Japan Earthquake, Seto et al. (2019:9) found that despite being expected to

show objective cost-effectiveness of their support activities, several organisations reported difficulties in producing evidence of effectiveness with figures
'"making it harder to obtain a budget'".

In April 2018, I visited one such *Kokoro no Kea* centre. The ordinary two-
storey concrete building was nestled tightly among a row of identical houses
on a side street off the main road, severely affected by the radioactive contamination from the damaged Fukushima Dai-Ichi Nuclear Power Plant. It
was only distinguishable from its neighbours by its clear glass automatic door
and the small sign above. It read: *Kokoro no Kea Senta*- Nozomi (Hope
Mental Health Clinic). Its muted green design was emblazoned with a pink,
anthropomorphised heart which was palely reflecting the setting winter sun.
Hardy, twisting plants crept through the cracks in the pavement and up the
rusty chain-link fence separating me from the train tracks across the street.

Crossing the threshold into the *genkan* (entrance hall), there was a narrow
gap for one to remove one's shoes and put on a pair of provided slippers. The
inside, however, felt spacious and bright. There was a hive of activity at the
far corner. Volunteers, consisting of local residents, many of whom had once
come to seek help themselves, were working busily, talking in hushed, hurried
tones. Sitting at the long table in the middle of the room, a group of men and
women visitors were chatting away. As I stepped through the door, I was
immediately met by Mr Watabe, the director of Kokoro Care Centre Nozomi
who is also a certified psychiatric nurse. He led me to a large open tatami
room upstairs scattered with objects, musical instruments, art equipment and
other knickknacks. Someone brought up some tea, and we watched it grow
cold as we talked. He told me stories about people who come to the centre
and how the centre helps them.

There was, for example, the story of a middle-aged woman who was afraid
of leaving her house and even inside her house was afraid to use tap water for
fear of radiation poisoning. She makes 'long irrelevant checklists' and spends
all day cleaning. Because she was unable to leave the house, he decided to
come by and bring her bottled water when he visited. She also has had trouble
sleeping and was prescribed sleeping pills by her doctor. Rather than try to
address each problem individually, he decided to support her holistically. It
wasn't what he had been taught, it was just something he had developed
through his work. Or the story of the man in his 70s whose house had been
in the evacuation zone. He now lived in what he called a 'garbage mansion',
single, living alone, and shunned by his relatives. The temporary housing
where he lived soon became piled up with garbage, leading to complaints
from his fellow residents. Watabe and his team helped clean as much as possible but decided it was best to leave what they could, as the elderly 'shouldn't
be forced to throw things away'. There was also the case of the man in his 50s
who was a self-confessed pachinko parlour addict (a gambling venue where
people win prizes that can be converted into cash), but before his habit could
even begin to be addressed, support workers realised that a more holistic
approach was necessary – support for alcoholism, personal hygiene, debt
management, family rejection, and self-esteem were needed.

Mr. Watabe told me the stress of the earthquake and nuclear accidents is not the cause of all mental illnesses. 'Among these, there are diseases that are caused by social backgrounds, such as anxiety disorders, depression, and alcoholism, and diseases that are difficult to cope with but not thought of as a specific mental illness.' With another person who had a drinking problem, there were many unintended benefits which may not have been accounted for but which he thought had a clear value to society and the individual. Whereas a standardised, dependency intervention is only counting the number of alcoholics who have stopped drinking over 'three months' rather than the number of alcoholics who have a home and have stopped drinking over a year (not counting factors which increase the likelihood of returning to alcoholism, such as homelessness), the ability for caseworkers to adapt proactively to meet difficult situations people face requires an imagination which cannot be measured in a standardised way. 'Still, some people are stressed by changes in the environment after being forced to abandon homes or the division of families, many of whom have symptoms due to stress', he reported.

Although *Kokoro no Kea* is an unregulated term many different kinds of organisations can adopt, the 'Fukushima Kokoro Care Centre' website estimates there were about 20,000 consultations between 2012 and 2014. Mr. Watabe said that funds to cover operating expenses for mental care are decreasing year by year, and there are concerns over whether sufficient measures can be taken for victims in need of care after services finally shut down for good in the coming years. He is worried that 'the need for psychiatric care will continue to increase rapidly after the earthquake. However, at present, the number of staff and the budget are insufficient'.

Since the nuclear accident, evacuation orders have gradually been lifted in Fukushima Prefecture. However, Mr. Watabe points out that returning to one's *furusato* (ancestral hometown), or moving to reconstructed public housing, means that people must move away from the community that was being formed in the temporary housing shelter. In areas where more than half of the returnees are senior citizens, anxiety due to a decline in economic activity also causes stress. 'Among the victims who lost their jobs due to the earthquake and nuclear accident, there are people who became alcoholics and alcoholics who resumed drinking after being evacuated', he said. 'Some people who have lost their role in society due to the earthquake and nuclear accident, sometimes turn to alcohol. If the community shifts in the future, more people will be stressed, and symptoms are expected to become apparent'.

Yamaguchi (2018:409) suggests *Kokoro no Kea* has contributed to raising public awareness of mental health issues based on the reasoning that anyone with *kokoro* could develop post-traumatic stress symptoms 'grounded in the recognition that *kokoro* (mind) can be hurt in the same way as a person's physical body'. However, she notes that '*Kokoro no Kea* has not necessarily succeeded in reducing the cultural stigma associated with mental illness or changing the general disinclination of Japanese people to seek treatment' (2018:407). For health care professionals, the term typically indicates services that identify and address pathological mental health issues, 'while community people usually expect hearty care, which may not necessarily include specialized or clinical care' (Suzuki et al., 2015a:2).

Consequently, 'the concept of *Kokoro no Kea* has failed to create a shared understanding as was revealed by the discussions among mental health care providers following the 2011 earthquake' (Yamaguchi 2018:409; see also Suzuki and Kim 2012:2). This is reflected in the overall definitional struggles over the terms 'mental health and psychosocial' which, as some have argued, are so 'vague as to permit any well-intentioned intervention, despite its consequences' (Aggarwal 2011:22).

Conclusion: Towards the Subjunctive in DMH

Therapeutic governance indicates the authoritative language through which people come to understand both the meaning of events and the potential suffering experienced or attributed to them (Pupavac 2002, 2005). It indicates that the majority of people exposed to disasters are not traumatised by events but that 'the problem' is often located in people's minds rather than as a shared social problem. Diagnostic frameworks like PTSD are exclusionary to the extent that not all people experience suffering in a way that conforms to standardised nosology, and it seems unclear to me whether psychological and psychiatric knowledge claims after disasters refer to a psychological reality independent of the conditions in which this knowledge is generated or, rather, are an epiphenomenon of practitioner and researcher data collection practices (Danziger 1994).

Despite calls to redouble MHPSS relief efforts after the disasters, the budgetary shortfall to provide it continues to grow. So-called survey fatigue in the form of excess health surveillance and monitoring is believed to have adverse effects on wellbeing, and practitioners are understandably reluctant to share anonymised patient data. A lack of standardised data available on psychosocial support interventions might be to blame for the perceived ineffectiveness of services (Tanisho et al., 2015). With *Kokoro no Kea* centres threatened by closure, individuals are increasingly reliant on the support of voluntary aid organisations (Yamaguchi 2018).

Psychometric research should also be questioned as a means to access the variables that influence how judgements about the catastrophic potential of different types of risk are formed, such as the perceived 'willingness to be exposed' of individuals affected by radiation contamination and their alleged decisions to either stay in their communities or move away. Mary Douglas's analysis of risk and blame, specifically the utility of her dyad of material and symbolic risk (Douglas 1966; Douglas and Wildavsky 1983; Douglas and Douglas 2002), is helpful here as whilst risk perceptions might be situated in specific contexts, knowledge about empirically 'measurable' perceptions about technological-scientific hazards is hardly morally neutral (such as for the mental health component of the Fukushima Health Management Survey). A further sociological criticism of the cognitive psychology approach to risk perception is that it 'examines risk from a static point of view as if it were a snapshot at a particular time' (Rogers 1997:745) outside the specific social context of an individual respondent's daily life (Wilkinson 2001:09). Considering the 'social life of risk' in Japan, the uneven uncertainty of radiation contamination produces its own dangers manifest in concerns over the psyche of the population and attendant concerns over (putative) anxiety and safety. Being more speculative than absolute, measurement of mental distress risk in the cases I examine carries different weight and meaning from one context of social life to another.

Designating populations as victims of trauma can conceal the ways in which 'experiences take on multiple meanings in a collective history, in a personal life story, and in a lived moment' (Fassin and Rechtman 2009:281). We need to move old debates in transcultural psychiatry forward in the field of DMH. As Kirmayer remarked, the classification of psychiatric disorders 'reflects largely American and European concepts of psychopathology based on implicit cultural conceptions of normality and deviance' (Kirmayer 1989:327), and Summerfield has raised concerns about reframing the suffering of war 'as a technical problem to which short-term technical solutions like counselling are applicable' (Summerfield 1999:149). Nevertheless, the cross-cultural validity of the trauma category is largely undisputed by practitioners as a 'universal reaction to severe stressors' (Osterman and de Jong 2007:439). We could instead locate in the sociology of commensuration processes new analyses of how different entities compare based on various metrics (Espeland and Stevens 1998) to see the shortcomings of the promise of scientific rationalism and quantification strategies. At the same time, by pointing out the irreducible black-boxing of experience into categories – making the incommensurable commensurable –creates yet another black box: 'experience'. Practitioners on the ground, however, are not necessarily stuck between the universal (diagnosis) and the particular (experience) precisely because they are not merely converting experience into 'raw' data.

Whilst a concept of commensurability offers room for an ethnographic approach to issues of comparison in evidence-based clinical guidelines, it does not offer much to explain how commensuration happens in practice. Using a subjunctive orientation to express how people can act without complete knowledge may enable us to understand the differences in approaching care in the *indicative* mood ('as is' or things 'as they are') against what Hardman and Ongoro call 'subjunctive' practice in medicine (things 'as if' or 'as they could be').

What becomes apparent are the benefits of promoting situations in which care services are improvised in a person-centred way. In comparison to research-focused initiatives which rely on extrapolated data to encode care to make it commensurable to global standards of evidence and large sample sizes, as Hardman says, we might wish to promote a change in mindset whereby clinicians can become 'more imaginative with how they approach people' (Hardman and Ongoro 2020:5). A complementary concept of subjunctivity expresses how people do things in practice, such as after commensuration processes have happened (particularly after a screening). This becomes an orientation towards the future possibility of recovery, treatment, reducing risk, and improvement in some way in the quality of life. In the present, clinicians are working towards a hoped-for qualitative change in the person's life (such as based on a person's self-reporting, 'I feel better today'). And 'by practising subjunctively, insofar as they focus on the co-construction of a temporary social world for a particular purpose, clinicians create the conditions for patients to "enact with" rather than be "acted on"' (Hardman and Ongoro 2020:5).

It should also be noted that commensuration processes are not happening synchronously. Different spheres of relations are involved at different steps. Not all clinicians are involved in the screening and involved in community welfare programmes. The data collectors may not also be the ones analysing the data. But the entire edifice is structured around the transformation of people into a countable population: victim, survivor, 'resident of temporary housing shelter', and so on. The second point about countable populations is the use of methods to enact realities. For example, the use of the K6 questionnaire to construct a version of distress in a way that can be fitted into wider policy implementation and into predictive models feeding into national imperatives at the population level and perhaps infer future economic productivity. Typically, organisations wishing to attract government funds for mental health initiatives will frame the intended benefits of their programme in a language that appeals to government productivity narratives. That is, ways in which risks of harm (to the state) will be minimised, such as either restoring or boosting health, preventing ill health, or reducing costs of state resources in some way.

In the subjunctive mode of medicine, action is deferred, or intention is cast backwards so that the 'reasons' why something is done are often exceeded by the capacity to explain them. In the visceral moments of interacting with those seeking support, what actually is happening is improvisation at the microlevel. But what resources, then, are people drawing on to achieve that? What is the hinterland of action in a given context? As a hazy space between fact and possibility, the subjunctive expresses situations hypothetical, contingent, imagined, expected, or hoped for. It allows us to talk sensibly about the counterfactual and the possible: 'what could be' and 'what might be'. I argue that in the clinical space of my research participants, subjunctive contemplation of the 'as-ifs' and 'what-ifs' of care works to situate my interlocutor's subjectivity and experience in relation to the past, present, and possible future, not only of those impacted but for themselves as well.

Implicit commensurability processes need to be made more explicit. Doing so would expose the gaps in current data collection and interpretation systems which are inadvertently removing the nuances of lived experience in favour of easier data management, on one hand, and potentially dubious forays into 'molecularised' health care, on the other. The idea of the subjunctive is an area I wish to explore further as we work towards the potential future for those affected by the Great East Japan Earthquake. I propose further scrutiny and examination is called for how mental health professions 'organise forms of suffering as bureaucratic categories and objects of technical intervention' (Kirmayer et al., 2010:170), particularly regarding validation of disease classification, by ensuring sensitivity to the context-specific aspects of local communities' multiple needs. Consumers of evidence ought to be aware of the inherent flaws in any system that attempts to quantify experience, particularly in the screening phase before interventions have been applied (in terms of including and excluding individuals based on needs-focused assessments). This includes any data metrics indices used in psychological measurements as well as structured diagnostic instruments. In addition, by taking the numbers to task, I argue it allows us to 'carefully assess the worlds that emerge alongside numbering practices and the ways that processes of governance work with and through numbers' (Holtrop 2018:12). Also at stake is the transformation of social ills into medical and psychological problems, for example, the definition of radiophobia amongst people living in proximity to the radiation affected areas.

Finally, further research and evidence are still needed to ascertain the most valid and relevant course of action in the production of positive mental health outcomes after disasters and establish methods for more flexible models of care to capture their value. The world is imperfectly knowable, although this does not mean it cannot be known. Commensurability frameworks at the moment are not simply flawed. Rather, they are a necessary condition or precondition for mental health care. Governing the uncertainty of life means that a reliance on structured tools to provide medical assistance is ultimately unavoidable. A subjunctive perspective could instead help us reimagine commensuration frameworks towards mutually desirable results, accepting that there is no guaranteed outcome.

References

Aggarwal, N. (2011). Defining mental health and psychosocial in the Inter-Agency Standing Committee guidelines: Constructive criticisms from psychiatry and anthropology. *Intervention*, *9*(1), 21–25.

Bolton, P., Tol, W., & Bass, J. (2009). Combining qualitative and quantitative research methods to support psychosocial and mental health programmes in complex emergencies [Editorial]. *Intervention: International Journal of Mental Health, Psychosocial Work & Counselling in Areas of Armed Conflict*, 7 (3), 181–186.

Bowers, M. E., & Yehuda, R. (2016). Intergenerational transmission of stress in humans. *Neuropsychopharmacology*, *41*(1), 232–244.

Breslau, J. (2000). Globalizing disaster trauma: Psychiatry, science, and culture after the Kobe earthquake. *Ethos*, *28*(2), 174–197.

Breslau, J. (2004). Introduction: Cultures of trauma: Anthropological views of post-traumatic stress disorder in international health. *Culture, Medicine and Psychiatry*, *28*(2), 113.

Csordas, T. J., & Harwood, A. (1994). *Embodiment and experience: The existential ground of culture and self* (Vol. 2): Cambridge University Press.

Danziger, K. (1994). *Constructing the subject: Historical origins of psychological research*. Cambridge University Press.

Doi, T. (1988). *The anatomy of self*. Tokyo: Kodansha International.

Douglas, M. (1966). *Purity and danger*. London: Routledge and Kegan Paul.

Douglas, M., & Wildavsky, A. (1983). *Risk and culture: An essay on the selection of technological and environmental dangers*. Berkeley: Univ of California Press.

Douglas, P. M., & Douglas, M. (2002). *Risk and blame: Essays in cultural theory*. London: Taylor & Francis.

Ecks, S. (2018). Biocommensurations. *Anthropology News, 59*(4), e162–e165.

Espeland, W. N., & Stevens, M. L. (1998). Commensuration as a social process. *Annual Review of Sociology, 24*(1), 313–343. doi:10.1146/annurev.soc.24.1.313

Fassin, D., & Rechtman, R. (2009). *The empire of trauma: An inquiry into the condition of victimhood*. Princeton, NJ: Princeton University Press.

Fernandes, B. S., Williams, L. M., Steiner, J., Leboyer, M., Carvalho, A. F., & Berk, M. (2017). The new field of 'precision psychiatry'. *BMC Medicine, 15*(1), 1–7.

Freidman, M. J., Keane, T. M., & Resick, P. A. (2007). *Handbook of PTSD: Science and practices*. New York: Guilford.

Fukunaga, H., & Kumakawa, H. (2015). Mental health crisis in Northeast Fukushima after the 2011 earthquake, tsunami and nuclear disaster. *The Tohoku Journal of Experimental Medicine, 237*(1), 41–43.

Garfin, D. R., Silver, R. C., Ugalde, F. J., Linn, H., & Inostroza, M. (2014). Exposure to rapid succession disasters: A study of residents at the epicenter of the Chilean Bío Bío earthquake. *Journal of Abnormal Psychology.*, 123(3), 545. doi:10.1037/a0037374

Gómez-Carrillo, A., Langlois-Thérien, T., & Kirmayer, L. J. (2018). Precision psychiatry—yes, but precisely what? *JAMA Psychiatry, 75*(12), 1302–1303.

Harding, C. (2017). Culture and psychotherapy in Japan. *The Lancet: Psychiatry, 4*(2), 102–103.

Hardman, D., & Ongoro, G. (2020). Subjunctive medicine: A manifesto. *Social Science & Medicine, 256*, 113039.

Holtrop, T. (2018). 6.15%: Taking numbers at interface value. *Science & Technology Studies, 31*(4), 75–88.

Kapferer, B. (2005). Situations, crisis, and the anthropology of the concrete: The contribution of Max Gluckman. *Social Analysis: The International Journal of Social and Cultural Practice, 49*(3), 85–122.

Karz, A., Reichstein, J., Yanagisawa, R., & Katz, C. L. (2014). Ongoing mental health concerns in post-3/11 Japan. *Annals of Global Health, 80*(2), 108–114.

Kato, Y., Uchida, H., & Mimura, M. (2012). Mental health and psychosocial support after the Great East Japan Earthquake. *The Keio Journal of Medicine, 61*(1), 15–22. doi:10.2302/kjm.61.15

Kim, Y. (2018). Seeing cages: Home confinement in early twentieth-century Japan. *The Journal of Asian Studies, 77*(3), 635–658.

Kirmayer, L. J. (1989). Cultural variations in the response to psychiatric disorders and emotional distress. *Social Science & Medicine, 29*(3), 327–339.

Kirmayer, L. J., Kienzler, H., Hamid Afana, A., & Pedersen, D. (2010). Trauma and disasters in social and cultural context. In: Morgan, C., & Bhugra, D. (Eds.), *Principles of social psychiatry*. New York: Wiley-Blackwell. pp. 155–177.

Law, J. (2004). *After method: Mess in social science research*. London: Psychology Press.

Law, J. (2009). Seeing like a survey. *Cultural Sociology, 3*(2), 239–256.

Mitchell, C. M., & Beals, J. (2011). The utility of the kessler screening scale for psychological distress (K6) in two American Indian communities. *Psychological Assessment, 23*(3), 752.

Moncrieff, J., & Timimi, S. (2013). The social and cultural construction of psychiatric knowledge: an analysis of NICE guidelines on depression and ADHD. *Anthropology & Medicine, 20*(1), 59–71.

Mosse, D. (2006). Anti-social anthropology? Objectivity, objection, and the ethnography of public policy and professional communities. *Journal of the Royal Anthropological Institute, 12*(4), 935–956.

Müller, R., Hanson, C., Hanson, M., Penkler, M., Samaras, G., Chiapperino, L. ... Latimer, J. (2017). The biosocial genome? Interdisciplinary perspectives on environmental epigenetics, health and society. *EMBO Reports, 18*(10), 1677–1682.

Niwa, S. (2014). [A new structure for mental health and welfare in the Soso area to promote the recovery of people in Fukushima from the 3.11 earthquake and nuclear power plant accident]. *Seishin Shinkeigaku Zasshi, 116*(7), 621–625.

Osterman, J. E., & de Jong, Joop T. V. M. (2007). Cultural issues and trauma. In T. M. K. Matthew J. Friedman, & Patricia A. Resick (Eds.), *Handbook of PTSD: Science and practice*. (pp. 425–446). New York: Guilford.

Pupavac, V. (2002). Pathologizing populations and colonizing minds: International psychosocial programs in Kosovo. *Alternatives, 27*(4), 489–511.

Pupavac, V. (2005). Human security and the rise of global therapeutic governance: Analysis. *Conflict, Security & Development, 5*(2), 161–181.

Raphael, B., & Maguire, P. (2009). Disaster mental health research: Past, present, and future. In S. G. Y. Neria, & F. H. Norris (Ed.), *Mental health and disasters* (pp. 7–28). Cambridge, UK: Cambridge University Press.

Rapp, R. (1999). *Testing women, testing the fetus: The social impact of amniocentesis in America*. London: Routledge.

Richard, G. F. (2012). *Lessons for Japan from U.S. rebalancing of mental health care. A report of the CSIS global health policy center*. Available at: https://ciaotest.cc.columbia.edu/wps/csis/0027238/f_0027238_26080.pdf [Accessed 12 January 2012].

Rogers, G. O. (1997). The dynamics of risk perception: How does perceived risk respond to risk events? *Risk Analysis, 17*(6), 745–757.

Rose, N. (1998). Governing risky individuals: The role of psychiatry in new regimes of control. *Psychiatry, Psychology and Law, 5*(2), 177–195. doi:10.1080/13218719809524933

Seto, M., Nemoto, H., Kobayashi, N., Kikuchi, S., Honda, N., ... Kim, Y. (2019). Post-disaster mental health and psychosocial support in the areas affected by the Great East Japan Earthquake: A qualitative study. *BMC Psychiatry, 19*(1), 1–13.

Setoya, Y. (2012). Overview of the Japanese mental health system. *International Journal of Mental Health, 41*(2), 3–18.

Shultz, J. M., Kelly, F., Forbes, D., Verdeli, H., Leon, G. R., Rosen, A., & Neria, Y. (2011). Triple threat trauma: Evidence-based mental health response for the 2011 Japan disaster. *Prehospital and Disaster Medicine, 26*(3), 141–145.

Stolz, R. (2018). Money and mercury: Environmental pollution and the limits of Japanese postwar democracy. *Positions Asia Critique, 26*(2), 243–264.

Summerfield, D. (1999). A critique of seven assumptions behind psychological trauma programmes in war-affected areas. *Social Science & Medicine, 48*(10), 1449–1462. doi:10.1016/S0277-9536(98)00450-X

Suzuki, Y., Fukasawa, M., Nakajima, S., Narisawa, T., Keiko, A., & Kim, Y. (2015a). Developing a consensus-based definition of "Kokoro-no Care" or mental health services and psychosocial support: Drawing from experiences of mental health professionals who responded to the great East Japan earthquake. *PLoS Currents, 7*. doi:10.1371/currents.dis.cfcbaf509711641ab5951535851e572e

Suzuki, Y., & Kim, Y. (2012). The great east Japan earthquake in 2011; toward sustainable mental health care system. *Epidemiology and Psychiatric Sciences*, *21*(1), 7–11.

Suzuki, Y., Yabe, H., Yasumura, S., Ohira, T., Niwa, S.-I., Ohtsuru, A., ... Abe, M. (2015b). Psychological distress and the perception of radiation risks: The Fukushima health management survey. *Bulletin of the World Health Organization*, *93*, 598–605.

Takahashi, Y., Yu, Z., Sakai, M., & Tomita, H. (2016). Linking activation of microglia and peripheral monocytic cells to the pathophysiology of psychiatric disorders. *Frontiers in Cellular Neuroscience*, *10*, 144.

Tanisho, Y., Smith, A., Sodeoka, T., & Murakami, H. (2015). *Post disaster mental health in Japan: Lessons and challenges*. Tokyo: Health and Global Policy Institute.

Tierney, K., (2015). Resilience and the neoliberal project: Discourses, critiques, practices—and Katrina. *American Behavioral Scientist*, *59*(10), 1327–1342.

Tol, W. A., Barbui, C., Galappatti, A., Silove, D., Betancourt, T. S., Souza, R., ... Van Ommeren, M. (2011). Mental health and psychosocial support in humanitarian settings: Linking practice and research. *The Lancet*, *378*(9802), 1581–1591.

Tol, W. A., & Van Ommeren, M. (2012). Evidence-based mental health and psychosocial support in humanitarian settings: Gaps and opportunities. *Evidence-Based Mental Health*, *15*(2), 25–26. doi:10.1136/ebmental-2012-100644

Wilkinson, I. (2001). Social theories of risk perception: At once indispensable and insufficient. *Current sociology*, *49*(1), 1–22.

Yamaguchi, S. (2018). Rethinking the concept of kokoro no kea (care for mind) for disaster victims in Japan. *International Journal of Culture and Mental Health*, *11*(4), 406–416.

Part III
Social Difference and Inequality

9 Japan's Gender Perspective after the Explosions at Fukushima Daiichi Nuclear Power Plant

Sunhee Lee

Introduction

I am a cultural anthropologist, and I have mainly been engaged in research and support activities for 'marriage-migrant' women in the areas affected by the 3.11 disaster. From the year following the disaster, I have been involved with empowerment projects for migrant women in Fukushima, alongside supporters from the Kansai region and Kanto region. However, while I have engaged with other migrant women, I cannot say that I have been actively involved with Japanese women in Fukushima Prefecture following the nuclear disaster. To put it directly, I did not know how to handle the heaviness, despair, and inexplicable feeling of tension that I felt each time I went to Fukushima after the disaster. I worried that I might unintentionally hurt those who were being thrown into confusion by all sorts of information, anxiety, dissatisfaction, and mistrust. However, I was able to work in a different way with migrant women, as a migrant myself. I began to observe that the remaining women often found sources of pride and energy by participating in disaster support activities, through new approval from others around them. This chapter recognises the complexity of different gendered experiences in the 3.11 disaster and argues that, despite recent developments in the discourse on gender in disasters in Japan, the nuclear disaster in Fukushima brought forth new forms of division and discrimination within the social orientation of communities. It further considers the gendering of voluntary evacuation in the context of entrenched gender roles and argues that there is a need to look at evacuation rights from more diverse perspectives. The sections that follow begin with my own story as a Korean woman who evacuated after the 3.11 disaster before proceeding to a literature review of gender in disasters and case examples from Fukushima.

Reflecting on My Own Experience in Relation to the Field

Sendai, where I live, is 100 kilometres away from Fukushima Daiichi Nuclear Power Plant. On March 11th, 2011, my home was in a state of great confusion. Many of my household possessions were damaged in the Great East Japan Earthquake, having fallen and broken during the intense shaking. My

DOI: 10.4324/9781003182665-12

mother, in her 70s, happened to be in Japan for her grandson's (my son's) primary school graduation ceremony and experienced the earthquake alone in a seventh-floor apartment. Fortunately, she was not injured, but she had elevated blood pressure and was having a difficult time when we entered an evacuation shelter. This was at the South Korean Consulate in Sendai, which served as a shelter. The following day, people began to rush in, and we were packed together like sardines, sleeping in a huddle at night.

Even before the earthquake, the original plan was that my family would return to Korea after my son's school graduation ceremony. After the earthquake, there was also my mother's health to think about, so our first course of action was to go to the consulate to request support to help us go home. At the time, all methods of transportation had stalled, gasoline could not be purchased, and there was no way to get to the airport. Like us, there were families who had planned to return to their home countries and couples who had just given birth who apparently had made the same request to the consulate. The next day, we were informed that the consulate would assist us in returning home and that they would accept advance requests for flights. At the time, only five or six families applied to return home. However, we then started hearing reports about the hydrogen explosion at the Fukushima Daiichi Nuclear Power Plant. From that point forwards, there was a flood of requests to return home and a lot of fuss about the order in which passengers could get on flights. I had hoped that we could get our turn sooner since my mother was elderly and could not speak Japanese, and I had a young daughter who was only five years old. However, our turn for a flight did not come for quite a while. We were just on the verge of giving up when we were finally contacted. We were able to return home to Korea at last, six days after the earthquake.

I saw unforgettable scenes as we left Japan to return to Korea. At the time, Niigata Airport was flooded with Chinese people. Most of them seemed to be young women who looked like technical intern trainees, as well as Chinese couples and children. Among them, I could see women who seemed to be leaving alone with their children. There was no one to see them off. At the departure gate of this airport full of Chinese people, I also saw a few white women leaving the country with their children. I saw a Japanese husband sending his white wife off with their child. For some reason, this scene of him kissing their cheeks, sending them off with a smile to reassure his worried wife and child, left an impression on me.

After returning to Sendai at the beginning of the next month, I learned many migrant women had been evacuating, without their Japanese husbands or families. In the area struck by the disaster, there was then a tacit pressure put on women who had left Japan and evacuated to their home countries temporarily. In my local area, there were rumours of resentment toward a wife who had 'run away' with her child, who later returned and pursued a divorce. I wondered what exactly was going on with the criticism, which was particularly directed at Asian migrant women. I began to see that this process was revealing a social stigma faced by married migrant women who, in a time of emergency, were not seen as individuals with the right to leave but only as 'somebody's wife'.

Now, ten years after the disaster, and with this chapter as a starting point, I am finally attempting to tackle gender issues after the nuclear disaster, in relation to both migrant and Japanese women. It is rather late for me to face this topic, but I have decided to use this chapter as an opportunity to approach the academic evidence and address my own questions on gender in this disaster. In doing so, I recognise that I write this chapter from an outsider position, as a Korean woman and a cultural anthropologist who has been living in Japan. Because of the ongoing COVID-19 pandemic, I have not been able to directly conduct interviews in Fukushima; instead, this chapter considers the nuclear disaster and gender-related issues through prior participant observation with migrant women in Fukushima, documentary review, and reviewing the existing research evidence on this topic. The following sections review general literature on gender in disasters before examining how gender perspectives have developed in relation to disasters in Japan. I then present case examples of forms of social divisions and discrimination that women have faced after the 3.11 disaster and ways that evacuation decisions were gendered by the media and in academic research. These examples from Fukushima show how women faced challenging circumstances; however, they also highlight where women have been change-makers and leaders in working towards disaster recovery and gender equity in Fukushima

Disasters and Gender

In recent years within disaster research, there has come to be acceptance of the idea that the degree of harm a person experiences through a disaster can depend on their gender – in other words, disaster risk is not gender-neutral. Gender came to receive attention in disaster research from the late 1990s, and in the 2000s, it was identified that women are more likely to be seriously affected by disasters than are men. In many countries, this is due to women facing limited access to the information and resources needed to assess disaster risks and prepare for, respond to and recover from disasters. In reflecting on international research on natural disasters and gender, Ikeda (2011: 74) identifies and presents specific vulnerabilities that women can experience in a disaster. First, there is a general trend that in many natural disasters, more women die than men. This gender disparity is also linked to socioeconomic status, as the higher the socioeconomic status of women generally is, the smaller the gender gap in the number of victims will be (Neumayer and Plümper, 2007). The second is that at the same time as women participate in an increasing amount of labour due to gender-based divisions, women tend to be at a disadvantage when it comes to accessing recovery resources – in other words, despite the increasing participation of women in the workforce, there are still gender disparities in access to recovery resources. The third is a trend for increased violence (including domestic violence) against women and girls during disasters, and the fourth is a disadvantage in that women are often excluded from public spaces for disaster prevention and recovery (Enarson and Morrow, 1998).

This type of vulnerability is not newly formed at the time of a disaster; rather, multiple studies have suggested that the vulnerabilities women face in disaster settings can be traced back to everyday (non-disaster) realities. These vulnerabilities become more striking at times of disaster. For example, Wisner and colleagues have described how disaster vulnerability and the capacity to respond to disasters are constructed to be gendered in communities. In other words, men and women do not have essential differences in disaster vulnerability from the start, and the vulnerability that can be seen in groups of women at times of disaster does not necessarily last forever. The circumstances of each group and individual – such as their gender, age, health, disability, class, ethnicity, and immigration status – are influenced by a variety of factors, and forms of vulnerability will depend on the community environment as well as the type of disaster (Wisner et al., 2004). Wisner and colleagues put forward the idea for gender mainstreaming, to reduce disaster-related risks through addressing possible inequities and discrimination that have come to exist within communities in everyday settings.

In disasters, the problem of leaving out women from disaster recovery plans has become clearer over time. Perez identifies issues with disasters and women in her book (Perez, 2020). One episode she makes clear is of an earthquake in 2001 in Gujarat, western India, which killed thousands and destroyed about 400,000 homes, leading to recovery efforts to construct new homes. One anecdote describes how houses without kitchens were built as a result of not even asking for women's opinions, let alone allowing them to participate in the project. The same thing happened in Sri Lanka four years later when the Indian Ocean tsunami hit the coastal areas. There is a history of failing to consider women in post-disaster recovery activities not only in low- and middle-income countries but in high-income countries as well, including the United States. The majority of the members of the 'We Will Rebuild' project, organised for recovery from the damage caused by Hurricane Andrew in the Gulf Coast (US) in 1992, were men. Perez highlights that planning for business centres, skyscrapers, and chamber of commerce facilities became the focus of conservative and narrow-minded white men in the group, leaving out basic necessities and community services, and letting those who were struggling in real life fall by the wayside. Then, reconstruction projects led by women were established, seemingly in order to cope with the effects of the male-dominated 'We Will Rebuild' project. Similarly, when Hurricane Katrina occurred in 2005, many women who lost their homes were low-income African American women; however, they were not considered in recovery planning. The public housing where many of these women lived was demolished, and as a result, women lost their former communities and were relocated to more inconvenient locations. Perez (2020: 320–327) suggests that all of this was the result of people in charge of planning reconstruction failing to consider women in low-income households and prioritising business interests in the redevelopment of the city.

In the midst of these identified gender inequalities, there has been a move from conventional models of disaster preparation, response and recovery, to additionally consider the importance of efforts to eliminate social inequalities. The importance of having a diverse range of people (including women) participating in policymaking has also come under new focus. At the same time, there has also been recognition of the agency of women to make changes (Enarson and Morrow, 1998). Experts in gender and disaster risk reduction (DRR) today are calling for women's empowerment and involvement as key actors in disaster recovery as well as the DRR decision-making process. This is because major disasters also serve as 'windows of opportunity' for overcoming disaster-related vulnerabilities. However, scholars have cautioned that if the underlying processes which perpetuate gender inequality remain unresolved, women may be used as 'props' for the purposes of DRR and development (Jones, 2005: 10–11). Underlying inequalities must be addressed in DRR, alongside efforts to empower women.

Disasters and Gender in Japan

The growing attention to gender issues in disaster research internationally has also had an effect on Japan. In particular, at the time of the Great Hanshin Earthquake (1995), the difference between the number of male and female victims became evidence that was used to expose a gender gap in Japan. There were 3680 women who died, compared to 2713 men (Gender Equality Bureau Cabinet Office, 2012). In the aftermath of this disaster, activists from Women's Net Kobe, which had been active three years prior to the disaster, played a central role in setting up a support network for women, distributing supplies, providing consultations for women over the telephone, and holding support seminars for women affected by the disaster. It is said that at the time, many consultations made over the telephone hotline concerned domestic violence. Then, the following year, this group collected records of the frank opinions of women immediately following the earthquake and published them as *Women Talk about the Great Hanshin Awaji Earthquake* (Women's Net Kobe, 1995).

In *Women Talk about the Great Hanshin Awaji Earthquake*, the hardships women faced are written in their own words. They reflected that those who felt they had suffered less damage compared to others felt unable to express their own fears; women received cold looks from people around them for applying make-up while living in evacuation shelters; public toilets at the time were unwelcoming to women, and there were problems with donation money for couples in common-law marriages. Women had to face these hardships in a social atmosphere which pressured them to just keep quiet and conform because it was a difficult time (Women's Net Kobe, 1995).

Thanks to the efforts of these women, various studies, practices, and proposals were created and carried out on natural disasters and gender in Japan following the Great Hanshin Awaji Earthquake. The Hyogo Framework for Action, adopted at the 2005 Second World Conference on Disaster Reduction in Kobe, incorporated the perspective of gender into the entire process of disaster preparedness. The framework underscored the importance of the participation of women in order to reduce vulnerability to natural threats and to build disaster-resistant countries and communities (Arai, 2012: 4). Based on subsequent experiences in the Niigata Chuetsu Earthquake, the perspective of disaster prevention came to be widely incorporated in Japan's Basic Plan for Gender Equality (Gender Equality Bureau Cabinet Office, 2010), and gender has come to be widely incorporated in Japan's Basic Disaster Management Plan.

Then came the Great East Japan Earthquake in 2011, followed by the tsunami and the nuclear disaster at Fukushima Daiichi Nuclear Power Plant. The National Gender Equality Bureau Cabinet Office immediately sent a document to organisations in disaster-affected areas about disaster response based on women's needs and childcare needs. Specific contents mentioned included (1) the inclusion of sanitary napkins, diapers, powdered milk, baby bottles, and the like in supplies provided; (2) setting up partitions for privacy, ensuring there were facilities in which women would not have to be concerned with men being able to see them (changing rooms, nursing rooms, bathing facilities, separate men's and women's restrooms), spaces for young children to play, including family areas, in evacuation shelters; and (3) reflecting women's perspectives, opinions and concerns in the management of evacuation centres and coordinating with local medical institutions, midwifery institutions, health centres, childcare and educational institutions, gender equality centres, and others (Arai, 2012: 16)

However, in the context of this unprecedented earthquake and the damage and turmoil that immediately followed, it is hard to think that anyone could have operated a shelter, which fully met gender-based needs or the needs of diverse groups of people. There were stories of women's underwear and sanitary napkins being sent back without ever arriving at a shelter, and one of my own acquaintances had to put up with substitutes for sanitary napkins during evacuation and then even had to undergo surgery due to uterine illness after the disaster. One shelter in Ishinomaki, Miyagi Prefecture, where I had been involved as a volunteer until early July (2011), had not been well partitioned to create spaces for women and did not even have a washing machine. Because of this, it was routine for the women in the shelter to wash their laundry by hand every day. I also had a friend who evacuated to her husband's parents' house instead of an evacuation shelter; however, she ended up having to look after a dozen relatives or so. On top of this, she had just had surgery for breast cancer.

As Uekusa and I mention in Chapter 10, there are still many insular local communities in disaster-affected parts of the Tōhoku Region, and people who hold patriarchal ideas from the past. There are still male-centred societies, where most heads of local government bodies are men, and most members of residents' associations (not to mention municipal and town councils) are also men. As a matter of course, the residents' associations and town councils of evacuation centres were also often men, and most members of steering committees were men as well, and there were many cases in which the needs of women, children, and minorities went unnoticed.

Under such circumstances, local and national women's groups began to assist women in shelters, with the idea that women should support other women. In Miyagi, there was an organisation that provided a service called 'Sentaku Net' for women who had trouble getting laundry done after the disaster, taking laundry on their behalf. For women who had evacuated to neighbouring municipalities because of tsunami damage in their home municipality, there are stories of women's groups at evacuation sites listening to their needs, giving hand massages, and distributing cosmetics. There were various efforts made for women affected by the disaster, including workshops for making small accessories. During life in evacuation, which seemed to drag on and on, gatherings called 'Ochakko Salon' (gathering to drink tea and chat) were held to help prevent evacuees from becoming isolated (Asano and Tendo, 2021).

In Fukushima, there were a great number of evacuees following the explosion at the nuclear power plant. At the time, 'Big Palette Fukushima', a convention facility in Fukushima Prefecture, accepted more than 2500 evacuees from areas affected by the nuclear disaster. For the first time in Japan, a public, specialised facility for women's support, a 'women-only space', was set up at the evacuation centre and operated until August 2011, when the evacuation centre closed. This 'women-only space' was called for by those in charge of operating the evacuation centre, who made a request to the Fukushima Gender Equality Centre. Women from three groups active in the area took responsibility for managing the 'women-only space', making it a reality. Mr. Amano, who was manager of operations of the centre at the time, has also been involved in the education of children with disabilities and worked at the prefectural Gender Equality Centre. Perhaps because he understood the need to consider women and minorities on a regular basis, he was able to keep an eye open for finer details, such as the 'women-only space' and the distribution of women's undergarments. He was also able to get his intentions up and running with real implementation thanks to his connections to the Gender Equality Centre and to the volunteers from local women's organisations (Amano, 2012). Solidarity and cooperation with this centre and these organisations from normal (non-disaster) times may have been an important reason for their success after the disaster.

While there is insufficient space in this chapter to mention all initiatives, the efforts of citizen groups and the Gender Equality Centres of local governments widely recognised the need to incorporate perspectives of gender and diversity during Japan's disaster prevention, mitigation, and recovery processes. These centres have been established in various prefectures and cities in Japan as regional bases for the realisation of a fair and equal society through gender equality, as focused on during the United Nations Decade for Women. Names vary from region to region: 'Gender Equality Centre' (男女共生センター) in used in Fukushima and Miyagi Prefectures, and 'Women's Centre' (女性セン ター) in Iwate Prefecture. Of course, there are still many issues to be solved regarding the effectiveness at the administrative level, as well as the participation of women and minorities in decision-making; however, there has been considerable progress in incorporating gender and diversity considerations into awareness of disaster prevention and response. Existing literature demonstrates the importance of women and diverse groups of people being able to come together and exchange opinions in normal (non-disaster) times, to best prepare for unpredictable disasters. Through opening up opportunities for women to take leadership roles in town development, they can also become leaders in reconstruction and restoration after a disaster. This perspective is not inconsistent with community disaster prevention as part of the Hyogo Framework. Rather, if a window of opportunity opens in the wake of a disaster for women and diverse groups of people to be able to participate in their local community, this can strengthen community disaster prevention for the next disaster, and they can be leaders in post-disaster reconstruction.

Divisions and Discrimination after the Nuclear Disaster

Despite the progress described earlier, it should be noted that most disaster and gender studies referenced in the preceding sections assume a 'natural' disaster. However, nuclear disasters add complexities to the disaster recovery process and can result in particular types of gendered issues, including social divisions and discrimination directed at women.

In Fukushima, once the 'safety myth' of nuclear power was destroyed by the disaster, there were also discrepancies in scientific assessments of damage from invisible radiation. Furthermore, there was a need for government and other authorities to work neutrally in order to help disaster survivors, and this resulted in many authorities having a vague stance on the disaster; it became a disaster of ambiguity where it was difficult to know what was happening. Many have pointed out that this led to 'divisions', in terms of different opinions on how to deal with the disagreements arising between groups of people (Ito and Kawamura, 2016; Sung, 2017; Yamashita, 2013).

Ito and Kawamura investigated the difficulties faced by 'voluntary evacuees' from Fukushima (i.e. those evacuating from areas other than the mandatory exclusion zone, including places labelled as 'voluntary evacuation' areas or areas with no evacuation orders). In their research, Ito and Kawamura present a structure of social divisions within Fukushima – which includes implicit considerations of gender – as follows: (1) division based on ideas about radiation, (2) division based on whether evacuation is deemed possible or not, and (3) division based on *one's husband's line of work* and living environment (Ito and Kawamura, 2016: 83). They suggest that each evacuee would make their own individual decisions while situated in these three divisions. However, another interpretation can be added to Ito and Kawamura's analysis, which would be 'feelings of remorse'. If those worried about radiation doses could not evacuate, they would feel guilty for risking their family's health; even if they could evacuate, they would receive criticism from evacuees who say they had abandoned their hometown and would still feel remorse. This became particularly gendered; above all, when it came to mothers and children who had evacuated, this group in particular would receive criticism from people around them and feel remorse for their families being separated. Mothers who behaved contrary to gender roles and contrary to the pressure imposed on them by society (e.g. to keep family together and/or to remain in their homeland) resulted in misunderstandings about their individual choices, and they experienced intense feelings of guilt. The experiences of mother/child evacuees are also presented in further detail in Chapter 13 by De Togni.

However, the source of division does not necessarily lie with those affected by the disaster but is rather a wider societal phenomenon. Voluntary evacuation has been treated as a matter of extremely personal preference rather than a right (Toda, 2016: 24). Yamashita has looked at the 3.11 disaster and argues that 'the culture of Japanese society originally must have been based on social sustainability, keeping individual desires in check, or rather, prioritising social sustainability even at the expense of the individual' (Yamashita, 2013). However, this valorises individual sacrifice for the community. Expectations of individual sacrifice become intertwined with gender norms and can result in particular pressure and guilt faced by women. Empirical research has illustrated the experiences of women after the nuclear disaster, who, in many cases, experienced an atmosphere in which they felt their individual choices were not respected. This is evident in the testimonies of many women:

> There are many people who hide the fact that they are going out for recreation because they are afraid of being criticised by the people around them. Many mothers living at evacuation sites hide the fact that they have evacuated, for fear of discrimination and bullying. A lot of people cover up different feelings to maintain a superficial peace.
>
> (Tanazawa, 2019: 24)

> People from Fukushima are conservative, aren't they? There are many people who think, 'the government says it's okay, so it's okay,' or 'NHK says it's okay, so it's okay,' and don't dig deeper even when pressed. There is always an air of *'no one should be giving their own different opinions'* [emphasis added].
>
> (Tanazawa, 2016: 38).

> [P]eople are evacuating even though the government says it's okay.
>
> (Tanazawa, 2016: 64)

> Even more than I want Fukushima to be restored, what I really want is for there to be a world where everyone can live in their own way, wherever they are.
>
> (Tanazawa, 2019: 205)

Izumi Yamaguchi (2013: 220), who wrote a book about the stories of six women who had evacuated from Fukushima to Okinawa, wrote of their actions: 'Here in Japanese society, full of the unreasonable peer pressure of mutual surveillance and regulation, choosing to evacuate yourself is nothing short of a battle requiring the highest degree of courage'. Reflecting on this courage, I would add that the true state of social divisions resulting from the nuclear power plant disaster is not the act itself of moving away and being separated from family and neighbours. The division is not created by these actions themselves. The reality of division lies in intolerance towards different views or different judgements when it comes to radiation. This underlying intolerance can also be displayed with particular intensity when certain social groups, in this case women (however, also migrant groups and ethnic minorities, as presented in Chapter 10), go against expected norms in decision making. The role of gender in wider views of evacuation is discussed further later, with a focus on how women became a focus of media reporting and academic studies in a 'gendering' of evacuation.

The 'Gendering' of Voluntary Evacuation

Beginning immediately after the earthquake, many researchers have investigated, analysed, and reported on evacuation and support for evacuees from the nuclear disaster. My particular concern is the gendering of nuclear damage, the gendering of voluntary evacuation, and the othering that women faced – and that these issues have received little recognition.

Regarding gender issues caused by the nuclear accident, Shimizu writes that it is women who are taking on a disproportionate burden of the environment being contaminated by radiation, as the bearers of children, and the people who tend to be responsible for domestic work and care-related labour such as childrearing and care for the elderly. Nevertheless, '[t]here are still not enough opportunities for the voices of women who have taken on these roles to be reflected in policy making. Therefore, it is necessary when thinking about nuclear weapons and nuclear power to consider not only peace, but gender-related issues as well' (Shimizu, 2016: 167–168).

This is an incredibly important point. What we know about the health effects of radiation is that children can be more sensitive to it than adults (UNSCEAR, 2013), and in a context in which there were disagreements between scientists about the effects of radiation levels in Fukushima (Leppold et al., 2016), there would have undoubtedly been heightened anxiety among mothers, contributing to voluntary evacuation. However, is children's exposure to radiation a problem only for mothers in the first place? Even if we set aside the problem that Japanese society still exists in a traditional system of a gendered division of roles in which men work and women reproduce, I am not convinced that there is a gender difference when it comes to awareness of children's exposure to radiation. In other words, this problem seems to have become gendered even though it is not a problem specific to women, especially mothers.

Ironically, it can also be said that researchers and the media have contributed to the gendering of voluntary evacuation. Although unintentional, much of the research related to voluntary evacuation has focused on mothers who evacuated with their children (Tanaka, 2016; Toda, 2016; Tsujiuchi, 2016; Yamaguchi, 2013; Tatsumi, 2017). These studies highlight the difficulties of mother–child evacuations, such as economic hardship due to living a double life, and the mental and emotional burdens brought on by separation from father figures. In many cases, they propose the support necessary to alleviate such anguish. As research, this should not pose any problems, but from the aspect of gender balance, it is biased.

The idea that voluntary evacuation equates to mother–child evacuation has been reproduced not only in academia but in the media as well. A serialised article in the Osaka morning edition of the Yomiuri Shimbun on March 2, 2012, titled 'Escape (3): Evacuees from Outside the Prefecture Drawing Near' mentions that 70% of evacuees who evacuated to Okinawa from outside the prefecture are mothers and children, and there is an interview with one mother who describes quarrelling with her husband, who remained in the area due to work. There are other articles, such as 'Mother–Child Evacuees Discuss Worries: Mother's Day Gathering of 25 in Aya, Miyazaki' in the Western morning edition on 14 May 2012, and 'Away from Home' in the 13 June 2012 Osaka morning edition, which included interviews of mothers who had evacuated with their children, bringing up the difficulties of evacuation. On 31 August of the same year, the *Tokyo Morning Edition* reported on the beginning of employment support for mother-child evacuees in Yamagata. From 7 to 8 September that year, the *Tokyo Morning Edition* serialised reports on mother–child evacuees, documenting their feelings of guilt over evacuating as well as their heartbroken states of mind. Newspaper articles like these have been published every year around 11 March, the date the Great East Japan Earthquake occurred.

Such a trend, though small in number, decreases the visibility of families who have voluntarily evacuated, as well as father–child evacuees. Content on fathers who have evacuated is scarce; however, one such example can be seen in an *Aera* article from 9 March 2020, by Yuya Kamoshita, who evacuated from Iwaki with his family. As a father who left his post at Fukushima College's National Institute of Technology, now a part-time lecturer in Tokyo to make a living, Kamoshita has taken a stand on his own against the national government and the Tokyo Electric Power Company (TEPCO) (Nomura, 2020). Although not all can act as Mr. Kamoshita has, there must be many fathers who want to protect their children from radiation. According to a questionnaire conducted by the Toho Area Research Institute in May 2012, about 20% of voluntary evacuee families included fathers (including those who only evacuated on the weekends). Even among additional comments written by those surveyed, there were many opinions from men who wanted to voluntarily evacuate because they were worried about radiation but could not due to work and financial concerns. Some wrote that at the very least, it would be better for their children not to be there.

The term *mother–child evacuation* evokes an image of mothers struggling and plays an effective part in appealing for the need for systematic change and support. However, the biased trend in research and news of 'voluntary evacuation = mother–child evacuation' erases the paternal awareness of the dangers of radiation and desire to protect children and has come to be accepted only as an issue of voluntary evacuees equating to mothers. Furthermore, it should be noted that there are people who are planting this mistaken preconception that voluntary evacuees only consist of nervous mothers and women who escaped, and who are utilising that preconception.

Maruyama, a philosopher, takes up poems composed by male poet Kenichi Iwai. Iwai wrote in response to poems by female poets who were also mother–child evacuees, about the uncertain and tragic evacuation of mothers and children. He writes:

> Awakening the radioactive egoism / frightening Tōhoku is being abandoned The self-sacrifice that should have been in Japan / 'as long as we are saved' As long as my child is safe/fanatical maternal instinct / deeply cold
>
> (Iwai, 2012)

In response to this way of thinking from Iwai, Maruyama (2015) points out that the 'crisis' symbolised by mother–child evacuation is first and foremost a family crisis as well as a community crisis, but it is also linked to a global ecosystem crisis. Maruyama highlights that the work of reproduction, the job of continuing life, is imposed on women, while men are expected to engage in the labour of production. Evacuation is situated within this wider gendered structure of society; only focusing on the mothers and children removes this important context and can result in evacuation being viewed as a problem of ignorant and sensitive mothers. There is a need for both academia and the media to be more gender-sensitive in this respect when reporting on evacuation after nuclear disasters.

Conclusion

At the end of March 2021, I visited three nuclear disaster-related facilities built in Fukushima: Commutan Fukushima in Miharu, the TEPCO Decommissioning Archive Centre in Tomioka, and the Great East Japan Earthquake and Nuclear Disaster Memorial Museum in Futaba. The cherry blossoms were in full bloom. It was a time when I reflected on the feelings of those who had to leave this beautiful place.

There were many forms of division and conflict that manifested after the nuclear disaster. People have tried to do what they can, under difficult circumstances, and there has been both discrimination directed at women and particular successes led by women. The director of the Iwate Gender Equality Centre recently reported on the experiences of women from the village of Iitate in Fukushima, which was completely evacuated under mandatory orders after the disaster. One can imagine the feeling of loss of having one of the most beautiful villages in Japan, a place built on the contributions of each villager, taken away from them in an instant. Still, even when they could not help but mourn, these women took to their feet. Chiba writes about women who were involved in the management of shelters and supporting their community, women who published their own personal histories, and women who restarted projects to create a network of female farmers and connect agriculture and food supply. Many of these women had studied in Europe as young wives in a village project 30 years ago, learning about various ways of life, and after returning to Japan they had expressed their opinions in the community and taken their roles as leaders (Chiba, 2019). Director Chiba suggests that it is important to strengthen people even before disaster strikes. It is important to have faith in your ability to make your own judgements and act in the event of an emergency. Therefore, it is necessary to have the power to think, the power to make decisions, and the power to help, even in normal settings outside of disasters. In any case, it is necessary to understand and acknowledge the status quo of diverse groups of people. There should be no differences in the importance of life between men and women.

After the disaster, various kinds of divisions arose. However, Fukushima Prefecture is currently making an appeal for revitalisation and holds high expectations for it. This chapter has highlighted the experiences of women and discussed gender issues in the wake of this disaster. There are limitations to this chapter in that it does not touch on male experiences, the full spectrum of gender, or gender minorities and the LGBTQIA+ community in Fukushima – instead, it primarily focuses on heterosexual women. This is, in part, because of the dearth of gender research conducted in this setting to date. My own expertise also relates to the experiences of these women. By connecting wider literature with the experiences of women and exploring their roles not only in relation to discrimination and challenging decisions but also as change-makers and leaders in Fukushima, it is my hope that this work can lead to further research that continues to advance gender studies in Japan.

References

Amano, K. (2012). あの時、避難所は・・"おだがいさま"が支えた169日間—福島県内最大級の避難所「ビッグパレットふくしま避難所」が教えてくれたこと— [At that Time, the Evacuation Shelter Was ... 169 Days Supported by 'Otagaisama'. Lessons Learned from 'Big Palette Fukushima', the Largest Evacuation Shelter in Fukushima] [online]. *Rikkyo University, ESD Laboratory*. Available at: https://rikkyo.repo.nii.ac.jp/?action=pages_view_main&active_action=repository_view_main_item_detail&item_id=6494&item_no=1&page_id=13&block_id=49 [Accessed 26 August 2021]

Arai, H. (2012). 災害・復興と男女共同参画 [Disaster, Reconstruction & Gender Equality]. In: Murata, A. (ed.) *復興に女性たちの声を* [*Women's Voices on Reconstruction*]. Tokyo: Waseda University Press: 1–20

Asano, F., Tendo, M. (2021). *災害女性学を作る* [Making Disaster Women's Studies]. Tokyo: Seikatsushisosha.

Chiba, E. (2019). 被災当事者の復興に向けた学びとエンパワメント [Learning & Empowerment for Recovery for Disaster Victims]. In: The Japan Society for the Study of Adult and Community Education (ed.) *The Great East Japan Earthquake and Social Education.* Tokyo: Toyokan Shuppansha.

Enarson, E., Morrow, B.H. (1998). *The Gendered Terrain of Disaster: Through Women's Eyes.* Westport: Praeger.

Gender Equality Bureau Cabinet Office. (2010). 第 3 次男女共同参画基本計画. [Japan's Basic Plan for Gender Equality, 3rd Edition]. 17 December, 2010. Available online: https://www.gender.go.jp/about_danjo/basic_plans/3rd/index.html [Accessed: 10 August, 2021]

Gender Equality Bureau Cabinet Office. (2012). 平成 24 年版男女共同参画白書 [2012 Version of the Gender Equality White Paper] [online]. Available at: https://www.gender.go.jp/about_danjo/whitepaper/h24/zentai/html/honpen/b1_s00_01.html [Accessed 26 August 2021]

Ikeda, K. (2011). バングラデシュにおける女性に対する暴力と「ジェンダーと開発」の展開: ある「草の根」女性運動家の語りから [Violence Against Women and 'Gender and Development' in Bangladesh: From a 'Grassroots' Feminist's Narrative]. *Bulletin of the Faculty of Education, Shizuoka University (Social and Natural Sciences and Liberal Arts Series 60).*

Ito, T., Kawamura, Y. (2016). 自主避難者の今何が困難を引き起こしているか——アンケート調査よりの分析 [What Is Causing Difficulty for Voluntary Evacuees Today: Analysis of Questionnaires]. In: Toda, N. (ed.) 福島原発事故漂流する自主避難者たち－実態調査からみた課題と社会的支援のあり方.[Voluntary Evacuees Drifting from the Fukushima Nuclear Accident: Issues & Social Support from the Perspective of Fact-Finding Surveys]. Tokyo: Akashi Shoten.

Iwai, K. (2012). 原子の死 [Death of Atom]. Kyoto: Seijisya.

Jones, R. (2005). Gender and Natural Disasters: Why We Should be Focusing on a Gender Perspective of the Tsunami Disaster. In: Asia-Japan Women's Resource Centre (ed). 女たちの 21 世紀 [Women's 21st Century] 42: 10–11

Leppold, C., Tanimoto, T., Tsubokurac, M. (2016). Public Health After a Nuclear Disaster: Beyond Radiation Risks. *Bulletin of the World Health Organization* v94(11): 859–860

Maruyama, T. (2015). 『母子避難』の悲劇性と持続可能社会への希求 [*The Tragedy of Mother-Child Evacuation & Hope for a Sustainable Society*]. *The Ryukoku University Philosophical Review* 29: 1–14

Nomura, S. (2020). 1億円受け取る世帯があれば、わずか8万円の人も…福島原発「強制避難者」と「自主避難者」に差 [If There are Households Receiving One Hundred Million Yen, there are Also People Receiving Only Eighty Thousand…. The Differences Between 'Mandatory Evacuees' And 'Voluntary Evacuees' from the Fukushima Power Plant] [online]. *AERA Dot.* March 7, 2020. Available at: https://dot.asahi.com/aera/2020030500017.html?page=1 [Accessed 10 August, 2021]

Neumayer, E., Plümper, T. (2007). The Gendered Nature of Natural Disasters: The Impact of Catastrophic Events on the Gender Gap in Life Expectancy, 1981–2002. *Annals of the American Association of Geographers* 97(3): 551–566.

Perez, C.C. (2020). *Invisible Women: Exposing Data Bias in a World Designed for Men* (edition Japanese). Tokyo: Kawade Shobo.

Shimizu, N. (2016). 核・原子力 話しにくい原発事故の被害 [Atomic/Nuclear Power: Difficult-to-Discuss Nuclear Accident Damage]. In: Kazama, T., Kaji, H., Kim, K. (eds). 教養としてのジェンダーと平和 [Gender & Peace as Culture]. Kyoto: Horitsu Bunkasha.

Sung, W. (2017). 原発事故後の生活変化とコミュニティ分断の実態 [Lifestyle Changes & Community Division After the Nuclear Accident] [online]. Available at: https://psych.or.jp/wp-content/uploads/2017/10/72-25-27.pdf [Accessed 26 August 2021]

Tanaka, S. (2016). 漂流する母子避難者の課題 [Challenges for Drifting Mother-Child Evacuees]. In: Toda, N. (ed). 福島原発事故漂流する自主避難者たち―実態調査からみた課題と社会的支援のあり方 [Voluntary Evacuees Drifting from the Fukushima Nuclear Accident: Issues & Social Support from the Perspective of Fact-Finding Surveys]. Tokyo: Akashi Shoten.

Tanazawa, A. (2016). 福島のお母さん、聞かせて、その小さな声を [Mothers of Fukushima, Let Us Hear Your Quiet Voices] Tokyo: Sairyusha.

Tanazawa, A. (2019). 福島のお母さん、いま、希望は見えますか? [Mothers of Fukushima, Can You See Any Hope Now?] Tokyo: Sairyusha.

Tatsumi, Y. (2017). On Qualification As Survivors and As Evacuees: Four Years of Mother-And-Child Displacement in Tokyo Due to Fukushima Daiichi Nuclear Power Plant Accident. *Journal of the Research Institute for Christian Culture*, Seisen University, 25: 65–81.

Toda, N. (2016). 放置できない自主避難者問題 [The Voluntary Evacuee Problem That Cannot be Left Alone]. In: Toda, N (ed.) 福島原発事故漂流する自主避難者たち―実態調査からみた課題と社会的支援のあり方 [Voluntary Evacuees Drifting from the Fukushima Nuclear Accident: Issues & Social Support from the Perspective of Fact-Finding Surveys]. Tokyo: Akashi Shoten.

Tsujiuchi, T. (2016). 大規模調査から見る自主避難者の特徴――「過剰な不安」ではなく「正当な心配」である [Characteristics of Voluntary Evacuees as Seen from Large-Scale Surveys: 'Reasonable Anxiety', Not 'Excessive Anxiety']. In: Toda, N (ed.) 福島原発事故漂流する自主避難者たち―実態調査からみた課題と社会的支援のあり方 [Voluntary Evacuees Drifting from the Fukushima Nuclear Accident: Issues & Social Support from the Perspective of Fact-Finding Surveys]. Tokyo: Akashi Shoten.

United Nations Scientific Committee on the Effects of Atomic Radiation, Sources and Effects of Ionizing Radiation (UNSCEAR) (2013). UNSCEAR 2013 REPORT Vol. I [online]. Available at: https://www.unscear.org/unscear/en/publications/2013_1.html [Accessed 26 August 2021]

Wisner, B., Blaikie, P., Cannon, T., Davis, I. (2004). *At Risk: Natural Hazards, People's Vulnerability, and Disasters*. London: Routledge.

Women's Net Kobe (1995). 女たちが語る阪神・淡路大震災 [Women Talk about the Great Hanshin Awaji Earthquake]. Tokyo: Mokuba Shoten.

Yamaguchi, I. (2013). 避難ママ―沖縄に放射能を逃れて [Evacuee Mothers: Escaping Radioactivity to Okinawa]. Tokyo: Aurora Jiyu Atelier.

Yamashita, Y. (2013). 原発避難――分断とシステム強化の狭間で (特集 生態危機とサステイナビリティ――フィールドからの アプローチ) [Evacuation of Nuclear Power Plants: Between Decoupling & System Enhancement (Special Feature: Ecological Crisis & Sustainability: Approach from the Field)]. Chiba: Ajiken World Trend: 33–36

10 Social Vulnerability and Inequality in Disasters

Marriage-Migrant Women's Experiences in the Tōhoku Region

Sunhee Lee and Shinya Uekusa

Introduction

In the midst of the COVID-19 pandemic, we have come to the tenth anniversary of the 2011 Great East Japan Earthquake, Tsunami and Fukushima Daiichi Nuclear Power Plant triple disaster (3.11 disaster hereafter). Disaster research on social vulnerability shows that certain groups of people in society are more likely to experience greater ('natural' and human-induced) hazard risks, and because of this, disaster impacts are often uneven (see, e.g., Cutter et al., 2003; Wisner et al., 2004 for general social vulnerability theory). Disasters are socially produced, and those who are marginalised by larger social forces such as patriarchy, economic inequality, (institutionalised) racism and various forms of social oppression often experience exacerbated disaster impacts and more difficult pathways to recovery (Tierney, 2014; Matthewman, 2015). In this chapter, we explore the last ten years of disaster and recovery experiences of Asian marriage-migrant women in rural Tōhoku who experienced positions of social vulnerability in relation to the 3.11 disasters. We present their lived experiences of disaster and examine the ways in which pre-existing social inequalities in the Tōhoku region impacted this particular social group during and after the disaster while also highlighting forms of resilience. Our findings suggest that the disaster experiences of marriage-migrant women were diverse: some were empowered due to the disaster and subsequent forms of altruism and new acceptance and support from communities, while others experienced powerlessness and isolation (Uekusa & Lee, 2020a, 2020b).

As some notes on terminology, *marriage-migrant* is a term commonly used in migration studies, particularly in Korea and Japan. In the context of Japan, this term typically refers to a migrant woman who marries a Japanese man, generally through an international marriage broker, and then moves to a rural area in Japan. The background to marriage-migration in rural Tōhoku is presented in further detail later. Second, there are key differences between migrants and immigrants. The term *immigrant* typically used when someone makes permanent or long-term move whereas *migrant* is used in more temporary situations. We do not intend to use the two terms interchangeably; however, marriage-migrant women often enter Japan on temporary visas, then

DOI: 10.4324/9781003182665-13

marry Japanese nationals and become permanent residents. Finally, the term *(im)migrant* is uncommon in Japan. The term *gaikokujin* (literally 'foreign national') is commonly used in Japan to describe anyone who does not hold Japanese citizenship and/or does not have typical Japanese phenotype. The authors opt for (im)migrants as the term *gaikokujin* or foreigner has an exclusive and sometimes racist connotation in Japan (see Torigoe, 2019).

Throughout this chapter, we make references to our own empirical studies and media/technical reports. From 2011 to 2015, we conducted 60 interviews with marriage-migrant women in the eastern coast of the Tōhoku region (including disaster-affected areas in Iwate, Miyagi and Fukushima Prefectures). Qualitative data was also collected from the last ten years of our ethnographic observations, as well as emails and phone (follow-up) conversations with marriage-migrants and local community organisations. The main author Sunhee Lee conducted extensive fieldwork (including ethnography and in-depth interviews) to document the narratives of marriage-migrant women in pre-disaster Tōhoku and the issues that they confronted. Because of her anthropological field research, Lee had personal and professional relationships with many marriage-migrant women in the region. After the 3.11 disaster, Lee continued conducting her ethnography while actively participating in disaster relief and recovery efforts to support migrant communities in Tōhoku. She continues to work with and for these marriage-migrant women, and most participants in our research were recruited through her personal and professional networks. To protect study participants' privacy, we have removed all identifiers and used pseudonyms in our analysis.

Contextualising Marriage-Migrant Women in the Tōhoku Region

Before contextualising marriage-migrant women in Tōhoku, we provide brief socio-geographic backgrounds of the Tōhoku region. The majority of disaster affected areas in Tōhoku – coastal areas of Iwate, Miyagi and Fukushima prefectures – is rural. With the exception of the city of Sendai (a city with a population of just over 1 million), these rural areas largely depend on primary industries. From before the 2011 disaster, there were significant issues of population decline, ageing and poverty in rural Tōhoku (Urano, 2014, p.773), and, in many ways, the 3.11 disaster exacerbated social issues and pre-existing social vulnerabilities. Akasaka (2013), who conducted anthropological studies in Tōhoku, noted that Tōhoku continued to exist as a domestic colony of Tokyo, producing and sending primary industry products to Tokyo and other urban areas (para. 6). Tōhoku is, and has always been, a poorer region, and regional disparity is a major driver of outmigration and depopulation. Akasaka (2013) continues to argue that, although Tōhoku is deemed to be a centre of the Japanese manufacturing, it is indeed 'not much more than a site for a subcontractor of a subcontractor of yet another subcontractor of a major company [such as Toyota and Sony]' (para. 10).

Looking to the wider state of socioeconomic inequalities in Japan, Osawa (2013) points out that even though the percentage of poor workers in Tōhoku is higher than other regions, the percentage of low-income *households* in northern Tōhoku does not appear as high. She explains that, in rural Tōhoku, even though individual workers tend to have low income, total household income may not appear as low because typically several income earners live in a household (Osawa, 2013, p. 35). In these multi-generational households, all household members are typically expected to work and financially contribute to make ends meet. This suggests that women typically take low-wage jobs (e.g., manufacturing and food-processing jobs) and are also expected to perform unpaid domestic and reproductive labour – a typical example of the 'double burden' that women may experience (see, e.g., Hoschchild, 1989). The intersection of such regional disparity and Japanese traditional patriarchy shapes the particular social experiences of women in Tōhoku. It is unsurprising that a trend was observed where young women tend to out-migrate from Tōhoku to the urban areas where they had (perceived) better life prospectus. This consequently results in a peculiar phenomenon known as the 'bride shortage' or 'bride deficit' in rural areas across Japan (Ando, 2009; Constable, 2009; Cheng & Choo, 2015).

The bride deficit was a serious community issue in these rural villages as (the oldest) sons of these farmer and fishermen families were struggling to find brides and have children to pass on their agricultural lands and family business. In rural ageing communities, single farmers also needed to find brides to care for their ageing parents (Lee, 2018). A solution to the bride deficit was to seek out brides from low- and middle-income countries in Asia. Sometimes local governments were actively involved in the recruitment (Le Bail, 2017). The increasing population of migrant women in Tōhoku over the last three decades has been associated with the growing number of brokered marriages between Asian wives and Japanese husbands (Lee, 2012b, p. 89).

Of the total number of international marriages between Asian migrant wives and Japanese husbands in Tōhoku (Statistics of Japan, 2019), 49.9% of these marriages in Iwate Prefecture were between Chinese women and Japanese men, 27.8% were between Filipino women and Japanese men and 12.5% were Korean women and Japanese men. In Miyagi Prefecture, 37% of international marriages were with Chinese women, and 36.9% were with Korean women. However, there were relatively fewer marriages between Filipino women and Japanese men in Miyagi. The pattern of international marriage in Fukushima looks different: 46% of international marriages were with Chinese women, and 31% were with Filipino women, while only 11.2% were with Korean women. While there are many Chinese immigrant women in all three prefectures, there is a different settlement pattern among marriage-migrant women. This is likely due to different marriage brokers linked to different local authorities and the 'chain migration' – migrants develop ethnic networks (i.e., their social capital) in a destination country which facilitates newcomers' migration and settlement process (see, e.g., MacDonald & MacDonald, 1964). However, the number of international marriages overall in Japan has declined since 2006, while, even after the 3.11 disaster, the number of international marriages remain at low yet stable rate (Figure 10.1).

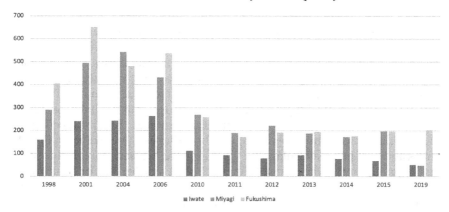

Figure 10.1 Number of international marriages in Tōhoku from 1998–2019.
(Source: Statistics of Japan, 2019)

As tends to be the case with Japanese brides, once a migrant wife becomes member of her Japanese family in-law, 'double duty' is expected, performing domestic and reproductive labour at home and work. Marriage-migrant women often work in low-paying jobs, typically in manufacturing or food-processing factories, where they do not need to communicate in Japanese language. Migrant women, just like Japanese women, face the traditional patriarchy and gender ideology that still remains present in many rural communities. However, for being Asian migrant women, their experiences are often further compounded by racial discrimination against minorities and the strong cultural and language ideologies. In general, marriage-migrant women face culture and language barriers yet feel stronger pressure to assimilate than other migrants (e.g., trainees and international students) because they are expected to be members of their Japanese families and become 'good wives' who learn and comply with the cultural expectations quickly. The Tōhoku region is relatively ethnically and racially homogeneous, and ethnic Japanese are the majority population. The percentage of foreign nationals in Tōhoku, including marriage-migrants, was significantly lower than the national average even prior to the disaster: 0.69% in Miyagi, 0.56% in Fukushima and 0.47% in Iwate (Ministry of Justice, 2010). Foreign nationals in Tōhoku are disproportionately concentrated in the urban areas such as the city of Sendai (i.e., international students and skilled migrants are exclusively concentrated in Sendai), and others such as trainees and marriage-migrants are widely scattered in the rural farming and fishing communities (Kamoto, 2014, p. 16). Without these migrant women, these declining rural communities would have remained relatively culturally and ethnically homogeneous, thereby these migrant women can be deemed as agents of 'internationalisation' in those socially and physically isolated monocultural rural villages.

Unlike typical migrants who tend to develop tight-knit migrant communities to help each other in destination countries (see, e.g., Portes, 1998), migrant wives in these already socially and physically isolated rural fishing and farm-ing communities often have very limited time and opportunities to go outside their communities and to interact with other migrants, except during Japanese-language classes (if they are enrolled) or local festivals. In this con-text, marriage-migrant women are expected to comply with the traditional, that is system – Japan's traditional family system (see Shimizu, 1987 for details) – and to deal with patriarchy, ethnocentrism, racism, language barri-ers, worker exploitation and economic disadvantages. The intersection of these social factors and geographic isolation shapes these women's 'double vulnerabilities' in a particular way. Furthermore, we have found that some marriage-migrant women came to marry Japanese fishermen or farmers for reasons including escaping from poverty in their home countries and search-ing for better life prospects in Japan. The fact that their visa and residency status in Japan depends on their spouses and their families (in-laws) can in some cases further contribute to disempowerment. It is then unsurprising that these women are less likely to speak out due to the language barriers and power imbalances, and domestic violence and abuse tend to be under-reported (Constable, 2012). As Parkinson (2019) notes, domestic violence against women often increases following major disruptions, including disas-ters. Under such circumstances, we have observed that, instead of speaking out and fighting for their human rights, many marriage-migrant women have learned to assimilate and invisibilise themselves as a 'strategy' to escape cen-sure, maintain their visa status and avoid racial profiling and other forms of oppression (Lee, 2013; Uekusa & Lee, 2020a; see Villegas, 2010 for strategic invisibilisation).

The 3.11 Disaster and Marriage-Migrant Women in Tōhoku

The pre-disaster invisibility and general social vulnerability of marriage-migrant women shaped their unique experiences of the 3.11 disaster. In the immediate aftermath of the disaster, many marriage-migrant women were unable to get the information in their languages and the support they needed due to the monolingual disaster communication (with limited information disseminated in English and grassroots-level efforts to translate and distribute information in multiple minority languages; see also Uekusa, 2019). More widely, the lack of information and fear of radiation created some panic among foreign nationals (including and beyond marriage-migrant women). Some quickly made their way back to home countries, taking the mercy buses and flights organised by the foreign embassies and consulates. There was a strong sense of solidarity to collectively respond to the historic disaster in Japan, and people – both Japanese and foreign nationals, who were rushing out, were generally labelled as 'disloyal' to Japan (Lee, 2012a). Many mar-riage-migrant women who left were particularly criticised for abandoning their Japanese families. *Sankei Shinbun* (2011) reported in an accusatory tone that marriage-migrant women with permanent residency status were leaving their children behind in Japan and returning to their home countries. This caused backlash. While there were some cases where this occurred, this was, for the most part, media disinformation and manipulation. From the very beginning, 'foreigners' (we purposefully use this term here), particularly marriage-migrant women, were stigmatised and distrusted in Japan due to the pre-existing forms of othering and homogeneous ideologies. The dark side to a strongly emphasised solidarity emerged across Japan after the 3.11 disaster.

As noted earlier, one of the main issues that these migrant wives (and other migrants) faced was the language barriers and information inequality, conceptualised as 'disaster linguicism' (Uekusa, 2019). It is critical to remember that many of these migrant women had very limited understanding of Japanese language and culture when they married and migrated to Tōhoku. They were typically provided very limited Japanese language education opportunities as national and local governments have no obligation to provide special assistance for migrants' social integration. Integration is outsourced, and migrants are rendered responsible. Some become competent in Japanese language over time. However, generally, there is limited Japanese-language proficiency in this group. In extreme situations, language can be a matter of life or death. Maria, a Filipino migrant wife in the Miyagi coastal area, told her friend's story:

> One of us perished in the tsunami ... well, the thing is that she must have not understood the tsunami warning announcement because the announcement was in Japanese, you know? And the announcement was using these words [such as *takadai* (high ground) and *hinan* (evacuation)] that are not being used in everyday life.

She asserts that, as her friend did not understand Japanese, the language barrier appears to be the reason for her friend's death. This might be an extreme case. However, empirical evidence of disaster linguicism abounds in disaster research. For example, Petri (2009) explains that, in the United States, 'Hurricane Katrina hit Latinos harder than other ethnic communities. Many Spanish-speaking Latinos did not evacuate – almost all storm warnings were broadcasted in English' (para. 4; see also Arlikatti et al., 2014). In the Tōhoku context, migrants were scattered across remote villages where very few authorities and community groups could provide multilingual services to start with, which made it difficult for marriage-migrant women to seek for appropriate support and to even speak out about their needs. Communication disparity and linguicism were, and still are, their everyday reality. This is particularly the case when communication is crucial for accessing critical information or expressing emotions. While Japanese speakers had more opportunities to exchange disaster information and discuss emotions of traumatic experiences at any time, with anyone, it was challenging for isolated marriage-migrant women who were novices in Japanese.

Another significant factor that exacerbated the social vulnerability of these individuals was their migration status and the lack of legal support. First and foremost, these migrant wives' (and, in some cases, their children who accompanied them from their countries of origin) visa and residence status as the

'Spouse or Child of Japanese National' were dependent on their Japanese husband and his family. In some cases that we have encountered, in-laws kept their passports so they would not be able to 'run away' (see Faier, 2008 for runaway brides). This created obvious power imbalances. Consequently, in the face of traumatic disaster event, some women – both Japanese and migrant – experienced domestic violence and emotional abuse at a higher rate (Yoshihama et al., 2013). During our interviews, it was common to hear that migrant wives were verbally abused by their husbands and family members, for example, with the following statements: 'Who's the one allowing you to stay in Japan?' 'If you run away, you'll be caught and deported!' 'Our children are Japanese, so you won't be able to take custody'. Legal support is available, however, often not accessible, or legal action is not a preferred option for women whose migration status depends on their husbands and in-laws. The lack of multilingual services (i.e., visa application and alien registration processes, disaster recovery information, disaster relief aid, information on radiation risks, and so on are available in Japanese) created situations in which these migrant women again needed to depend on their spouses and families, even in abusive situations. Intertwined with the pre-existing gender hierarchy, linguicism and racial discrimination against Asian minorities/migrants, these vulnerabilities only heightened during/after the 3.11 disaster.

As briefly noted, these examples show how marriage-migrant women in this particular Tōhoku context face highly complex forms of disaster vulnerability attributable to various forms of social oppression. The pre-existing social disadvantage of being Asian migrant wives in socially and physically isolated, traditional Japanese communities, who had limited cultural and linguistic competency, shaped vulnerability during the disaster. Therefore, we must be mindful that, as the vulnerabilities of marriage-migrant women in Tōhoku context were results of the intersection of social forces such as patriarchy, linguicism, institutionalised racism (i.e., legal/migration system), monolingual and homogeneous ideologies, traditional Japanese family system, seniority and so on, it is highly troubling to expect that their self-help, resilience and empowerment can solve complex issues without properly addressing social structural issues (Uekusa & Lee, 2020b). Simply put, migrant women should not be responsible for their own fate and empowerment without all cultural, political and social level efforts to address and solve the macro-structural issues such as racism and gender inequality. Their post-disaster experiences, however, show some concrete examples of what has helped or hindered empowerment (see also Uekusa & Lee, 2020b). The following sections provide further detail on the remarkable resilience of marriage-migrant women, which helps us move towards more realistic and constructive discussions beyond the usual discussion of labelling them *either* vulnerable or resilient.

Key Issues in the Long-Term Recovery

After the 3.11 disasters, the pre-existing social inequalities mentioned above contributed to further deteriorated economic situations and further social isolation for many marriage-migrant women. After the initial disaster response period, many researchers, practitioners and supporting organisations tried to give voice to these migrant women and identify points of vulnerability and support needs. However, invisibility and (in)voluntary isolation meant that it was difficult for government officials, researchers and supporting organisations to reach them. In the cities of Ishinomaki and Kesennuma, survey studies were conducted in 2012 and 2013 by the Support Centre for Foreign Nationals – an organisation supporting migrants in the area partnered with the local authorities – to understand migrants' specific disaster vulnerabilities (Support Centre for Foreign Nationals, 2012; Support Project for Foreign National Disaster Victims, n.d.). Findings revealed that the language barriers, a lack of employment and deteriorated family relationships were particular concerns among the respondents. In this section, we discuss some of the hardships experienced by marriage-migrant women in the long-term recovery period.

As with other disasters including the current COVID-19 pandemic, the 3.11 disaster exposed otherwise invisible inequalities in Tōhoku. It also exposed otherwise invisible migrant women in rural Tōhoku. The disaster ironically provided an opportunity to expose and address the root causes of social vulnerability such as economic disadvantage, patriarchy and racism. However, it was difficult for researchers and practitioners, who were 'outsiders' to these rural communities, to reach out to foreign nationals in the area and to even confirm their safety (Minagawa, 2012). Foreign embassies needed to confirm the safety of their nationals. However, because many marriage-migrant women, especially from China and Korea, were not using their real names (as one invisibilising and integration strategy), they were extremely difficult to reach (see Lee, 2013; Uekusa & Lee, 2020a). In some cases, their Japanese in-laws did not even know their real names. Needless to say, those who looked Japanese and had already become Japanese citizens were very invisible to outsiders including researchers, government officials, lawyers, journalists and supporting organisations. The foreign national support organisations ended up relying on migrants' own ethnic networks. Even so, some extremely socially and physically isolated migrant women are thought to remain isolated. Thus, despite significant efforts to reach out to as many these migrant women in the disaster affected areas as possible, we may not know the whole picture of this particular group yet. Furthermore, while some large-scale studies have been conducted (as presented earlier), we suggest that survey-based studies have generally not sufficient to give voices to these migrant women in the area. Further qualitative work is recommended to understand their experiences of the 3.11 disaster, its recovery and the 'new normal' for marriage-migrant women.

Disaster linguicism not only caused information and communication inequality in the wake of the disaster (see Uekusa, 2019) but also contributed very limited employment opportunities. In the disaster affected areas, there were higher demands for physical labour such that men tended to find disaster recovery-related temporary jobs more easily. There were, however, many fewer *paying* job opportunities for women, even though women were expected to provide free labour (i.e., women needed to perform unpaid domestic responsibilities and reproductive and community labour, for example at evacuation centres, due to certain cultural practices and increased gender norms; see also Enarson & Fordham, 2001; Enarson et al., 2007 for discussion of this phenomenon in other settings). This is highly problematic for marriage-migrant households. On average, household economic wellbeing is often relatively poor to start with because the husbands of marriage-migrant women tend to have low paid jobs. Therefore, these households were already at higher risk of poverty, particularly in the face of crises. Migrant wives needed to perform dual duties to keep their families afloat; however, due to limited Japanese language proficiency and other possible obstacles such as racism, we found that it was extremely challenging for them to find new jobs when manufacturing, food processing and other jobs which do not require high language competency, were simply not available during this period.

Furthermore, for some, family relationships were negatively affected after the disaster. One Korean marriage-migrant women, Jina, in her 40s moved into a temporary housing site with her family after the whole area where she used to live was inundated by the tsunami. She lived with her husband, their four-year-old son, and her mother-in-law, in a tiny temporary unit. After an argument with her mother-in-law, the mother-in-law left the family and moved into her daughter's house in Tokyo. Since then, Jina explained during our interview, she felt very excluded from the small temporary housing community because she was unreasonably blamed for being 'unkind' to her mother-in-law and not being 'good wife'. For Jina, the disaster ironically provided an opportunity to get out of her house and network with minority groups and disaster relief groups. Jina invited a group of disaster evacuees from her temporary housing community to make small accessories for fundraising, and this charitable activity was going well. She felt included and empowered. However, after the incident with her mother-in-law and consequent backlash, her relationships with her neighbours in the community deteriorated. For nearly a year afterwards, she lived in her tiny temporary unit, avoiding neighbours.

As noted earlier, disasters often expose inequalities in society, but the 3.11 disaster also exposed otherwise invisible migrant women in rural Tōhoku to the wider public. The formerly invisible migrant wives' remarkable resilience and loyalty to Japan and their Japanese families grabbed media and public attention, which, in some ways, interacted (for better and for worse) with ways that existing vulnerabilities were socially produced for this group. Positives and negatives came out from their increased visibility or, as Leong et al. (2007) call it, 'hypervisibility'. We previously argued (Uekusa and Lee 2020a, 2020b), using Brighenti's visibility theory (2007), that unfortunately those who became hypervisible or went over the perceived visibility threshold during/after the disaster, tended to become targets of backlash. Stigmatisation of visible migrants were seen at all levels. As the promotion of multiculturalism and response to multilingual demands in disaster recovery and governance, the city of Ishinomaki hosted a public hearing in February 2014 about coexistence and multiculturalism in their city and drafted the Ishinomaki Plan for the Promotion of Multiculturalism (Ishinomaki City, n.d.). Public opinions about the plan were mixed. Some were supportive, but the anti-migrant sentiment and Japanese homogeneous ideology seemed to dominate the debate. Amongst the negative opinions, one stated: 'Ishinomaki should focus on own citizens', 'Foreigners are robbers and scary, and we're concerned about our community security'. Some Japanese families might have been jealous of these migrant women being in the 'spotlight' while evacuees, especially older adult evacuees, often remained invisible to the wider public. Public attention, volunteer groups and relief aids seemed unevenly distributed, depending on how 'visible' certain groups or areas were, often through media (Lee et al., 2015). Like Jina, some of the study participants also noted that their relationships with their in-laws worsened for various reasons during this time, which caused further isolation, poverty and poor mental health outcomes (Lee, 2012b; Uekusa & Lee, 2020a).

Grass-roots Efforts to Support Migrant Communities

There were some positives that came out of the increased visibility of migrant women. Due to disaster altruism (see, e.g., Solnit, 2009; Matthewman, 2015), the 'visible' resilience and vulnerability of migrant women in disasters featured (and sometimes was speculated about) in media. In some cases, this helped these migrant women connect with other members of their ethnic community and support groups (Leong et al., 2007; Lee et al., 2015; Lee, 2018; Uekusa & Lee, 2020a, 2020b). As described earlier, some of the most pressing needs of migrant women after the 3.11 disaster were employment support and Japanese-language learning opportunities, and many support groups heard these requests and started focusing their support on these issues. In particular, assistance for the Nursing Care Certificate Level 2 (*kaigo in 2 kyū yousei kenshu katei*) was in high demand among migrant women, particularly Filipinos, and many volunteer groups organised some courses for them. This happened throughout the disaster-affected area and helped improve employment opportunities by opening up path to certification to work in nursing homes.

In Kesennuma (Miyagi Prefecture), a Japanese language class was funded by the Japan Association for Refugees (JAR), in cooperation with local Japanese-language teachers, for those who were studying towards a nursing care certificate. In its first term from June 2011 to March 2012, the class helped nine Filipino women to qualify. After that, the same course was held from July 2012 to January 2013, and 15 Filipino and Chinese marriage-migrant women obtained their certificates (JAR, 2011). From July 2014 to January 2015, the Let Us Walk Together! Project, a project of an Anglican church association, Nippon Sei Ko Kai (NSKK), for supporting disaster survivors, held a course and helped a total of 31 migrant women to become certified caregivers (NSKK, n.d.). These volunteer activities and its positive impacts on Filipino women in Kesennuma were particularly covered in media, which helped them further extend their 'disaster social capital' to leverage more resources (see Uekusa et al., 2022 for disaster social capital).

In Minamisanriku (Miyagi Prefecture), the NSKK began the nursing care certification course for Filipino women in July 2011. A prefabricated building was set up for Filipino women community leaders in the town to provide a Japanese language course, including childcare for those who were mothers. As a result, six migrant women in this town became certified nursing caregivers in April 2012. However, because this town was severely damaged by the tsunami, there were still no operating nursing care facilities yet where they wished to acquire caregiving jobs. Furthermore, a Filipino community leader, Erika, spoke about the general backlash of migrants providing paid caregiving work and being in the spotlight: 'There were some cases in which Japanese colleagues became jealous about migrant caregivers. Filipino caregivers were then assigned harder and less desired tasks such as wiping care recipients' body' (see also Satake et al., 2015).

There were also faith groups in Sendai which set up the Support Centre for Foreign Nationals in Ofunato (Iwate Prefecture) in November 2011, and two priests from the Philippines and Indonesia joined to assist the group. At this support centre, a nursing care certificate class was also offered; seven marriage-migrant women became certified from the first cohort, eight from the second cohort and eight from the cohort. In Watari, a small coastal town in Miyagi, 15 migrant women attended a nursing care course. However, the priest in charge of the program said that there were now only three women in Ofunato and three in Rikuzentakata who were actually working as caregivers. Many certified caregivers were working in the recovery process instead, sorting out the debris, which paid better wages. Caregiving is physically and psychologically demanding, yet the wage is considerably low. This priest further explained us that, in 2014, the tsunami-damaged seafood-processing factories started reopening, and many migrant women, who were certified caregivers, decided to return to these food-processing factories. Tsuchida (2016) notes that supporting groups need to understand the context of migrant women's experiences. It is important to strengthen these women's human capital, yet focusing solely on their work skills do not solve the structural issues that they confront.

Despite some failures and unintended consequences, these external support organisations have provided a turning point in the lives of some migrant women. With support, these migrant women became very active in the disaster-affected areas, improving their Japanese-language competency and increasing their human capital. These grass-roots efforts are particularly important because, while the Japanese government offer 'highly skilled migrants' more support, these who migrated to Japan for marriage were left without any official support. These external organisations addressed immediate and specific needs of migrant women, in an 'outsourcing' of support systems (away from government), dependent on volunteers and grass-roots efforts.

Self-Organising and Empowerment

A key change after the initial external support was that marriage-migrant women initiated their own organisations. Until this point, migrant women were more loosely connected in the Tōhoku Region. However, after the 3.11 disaster, migrant women, within the Filipino community in particular, have self-organised (and extended their existing networks) in what can be seen as a form of self-empowerment. These newly emerged and extended networks functioned as bridges between volunteers/supporting organisations and the disaster victims such that relief aids, donations and information about recovery and employment were distributed to the otherwise difficult to reach migrant women in remote areas. Filipino women's self-organisation and empowerment inspired other migrant women, who later developed their own networks for various purposes. Table 10.1 lists (some) examples of migrant women's emergent and extended community groups in Iwate, Miyagi and Fukushima.

In Fukushima, partly due to the mass evacuation and difficulty to organise, there was no external support for migrant women in the immediate aftermath of the nuclear disaster. Therefore, unlike Iwate and Miyagi, marriage-migrant women in Fukushima self-organised from the start and initiated activities to help each other even in the absence of wider support. Initially, Filipino migrant women started community groups in the cities of Fukushima and Shirakawa. Chinese migrant women formed groups in Sukagawa, Iwaki and Koriyama. In the face of radiation risks, these women were particularly concerned about their children's health and supporting their households' financial wellbeing.

A Filipino community in Fukushima, Hawak Kamay (holding hands) Fukushima (HKF), was started by Jasmine, a volunteer at the Fukushima International Association. As she told us, Jasmine is a Filipino marriage-migrant woman who, through her marriage, migrated to Japan in 1989. Although she had been in Japan for more than 20 years, she had never really met and associated with other Filipinos in the area. After the disaster, the Philippine Consulate contacted her and asked her to receive and distribute relief supplies to other Filipinos. She explained us that she then tried to find and contact Filipinos in the local area through mutual friends and word of mouth. This was how Jasmine and other Filipino women started self-organising and created a community group in Fukushima. In the face of the triple disaster, these Filipino women had common hardships and developed mutual support: they visited temporary housing sites where Fukushima residents were evacuated due to the radiation risks. This Filipino group prepared food from the Philippines and cheered people up with the Filipino songs and dances as a form of cultural exchange in temporary housing. Jasmine stressed, '[W]hen I saw Fukushima people evacuating and living in the evacuation shelters, I thought that they were just like us, *migrants*'.

HKF's activities covered the local area and were featured in media reports (see, e.g., *Mainichi Shinbun*, 2017). Jasmine's daughter, who did not previously positively identify with her Filipino descent, wrote an essay about her origin and how she is proud of who she is as a person of mixed race. Her essay won the Human Rights Award at the time. 'I felt like I was part of this Japanese society for the first time', said Jasmine. In the face of extreme situations, people can become altruistic and selflessly help each other, and this can lead to situations where everyone is cared for and included (Matthewman & Uekusa, 2021; Solnit, 2009). Jasmine shows a case in point: migrants felt accepted and become part of the disaster *communitas* (Matthewman, 2015).

The preceding examples were not limited to Filipino women but were seen widely from marriage-migrant women from different countries of origin. Chinese women also self-organised and developed community groups to help each other and others in need, especially in Fukushima. As with HKF, marriage-migrant women realised the need for their own ethnic community when they helped the embassies and consulates confirm the safety of their nationals. There were many Chinese migrant women married to Japanese men in Fukushima, and Chinese-language classes for their children was a community interest. Chinese-language classes were organised first in Sukagawa, then in Iwaki and Koriyama. All the community leaders who organised and ran the classes were experienced Chinese teachers in their local communities. Support was provided by local Japanese communities and Japanese families through their husbands. At first, this group had to create all teaching/studying materials; however, with the cooperation of the Chinese Consulate, they were able to leverage resources and acquire study materials for students free of charge.

The community groups formed by migrant women after the 3.11 disaster offered various social functions and services ranging from: teaching their children and families their heritage languages, legal and general consultation groups and so on. Overall, these networks and their critical role in disaster response led to empowerment. With this bottom-up grass-roots effort, there were also cases where local governments helped in promoting multiculturalism in Fukushima. In the wake of the 3.11 disaster, migrant women in Tōhoku, who had been invisible, isolated and excluded for a long time, now were able to speak out and feel like they were active members of a community. This can be a major positive coming out of the devastating triple disaster.

Disaster Social Capital and Migrant Women's (Un)sustainable Empowerment

As noted earlier, a major positive change that the 3.11 disaster brought to migrant women communities was their new and extended social capital (and resources and support leveraged through that). These emergent networks temporarily played a critical role in filling the gaps in social and human services (see Uekusa et al., 2022). Similar examples were seen in previous disasters, including the case of Vietnamese- and Spanish-speaking communities following the Great Hanshin Awaji Earthquake (Yoshitomi, 2008), and the case of the Vietnamese community in New Orleans after Hurricane Katrina (Leong et al., 2007). The 3.11 disaster caused devastating social and physical damage; however, some of the formerly invisible and isolated marriage-migrant women experienced empowerment and felt included for the first time as active members of their communities and wider Japanese society. However, we need to be cautious here. Their disaster social capital and empowerment were, in many cases, linked to external factors, especially the external support groups who brought the financial and logistic needs from the outside, particularly in Miyagi (Uekusa & Lee, 2020b). This 'linking social capital' (see Aldrich, 2012 for different types of social capital) was a critical contributing factor to the (un)sustainability of migrant women's community empowerment and activities in Tōhoku. Furthermore, when migrant women did not have the support of their local communities (and Japanese families), their newly hypervisible status caused further backlash from some local Japanese people who were also socially isolated and marginalised (see also Uekusa & Lee, 2020b) and meant that this empowerment was unsustainable.

To elaborate the preceding points, we can look back at the nursing care training programs which were meant to increase migrant women's human capital and the local communities and families who actively supported this. In Kesennuma, JAR mainly funded the program, and local municipalities, private sectors, language teachers and other community members also assisted. Such a collaborative effort was deemed to be the key for success (i.e., some local nursing homes supported the program by providing job opportunity after they became certified). In other areas where the government and local support was absent, it was difficult for the certified caregivers to find a job.

In some cases, the employment support programs, developed by 'outsiders' who may have overlooked the larger social forces when trying to help migrant women's empowerment, have caused some unintended consequences. For instance, the Social Enterprise English Language School program was initiated to train Filipinos to be English-language teachers. This particular employment support program, was started by Filipino missionaries, provided Filipino women with specialised English teacher education. Many Filipino women became part of this program as they wanted to achieve upward social mobility by becoming English teachers. However, in the market, there was a strong demand and preference for native English speakers with possibly typical Caucasian phenotypes. In some cases, these Filipino teachers started a private English conversation class in their local community; however, because the teachers were not native speakers, they were turned down by public schools. At HKF in Fukushima, they started an international preschool for their own children, so these trained English teachers can teach children English. However, due to the lack of enrolments and stable income, the classes were quickly closed at the end of 2012. In addition to issues of discrimination in preferences for English teachers, the community leaders also needed to deal with the debts such as rent and instructor salary. Many of these Filipino women dreamed of becoming English teachers in Japan, yet their empowerment was blocked due to structural issues which were beyond their control.

Migrant women's empowerment and active community participation must be supported by the local community and wider public, and the issue of power has not yet been properly addressed and resolved. Some of the community groups created by migrant women in Tōhoku (shown in Table 10.1) have already been suspended for various reasons. For example, Takako, who came to Japan in 1999 from Korea for marriage, started and was the chairperson for an non-profit organisation (NPO), which had promoted multiculturalism, offered consultancy and hosted cultural exchange events in the region even prior to the 3.11 disaster. She said at the multicultural coexistence symposium hosted by Miyagi Gakuin Women's University in December 2011:

Table 10.1 Migrant women's community groups (emergent and pre-existing) in Iwate, Miyagi and Fukushima (as of December 2017)

Region	Organisation Name	Ethnicity	Main Activities	Established
Ofunato/ Rikuzentakata, Iwate	Pagasa	Filipino	Consultation and support for Filipino living in the area. Interaction with the local community	November 2013
Kesennuma, Miyagi	Bayanihan Kesennuma Filipino Community	Filipino	FM radio program, cultural exchange, safety confirmation, fellowship, etc.	May 2011
	Kabayan Kesennuma Filipino Community	Filipino	FM radio program, regional cultural exchange	March 2014
Minami-sanriku, Miyagi	Sampaguita Fighting Ladies	Filipino	Fellowship	July 2011
Ishinomaki, Miyagi	Happy Mothers Club	Multiethnic	Parental support for migrant mothers, cultural exchange	
	International Global Village non-profit organisation	Multiethnic (Korean)	Consultation for migrants, Japanese language classes, promoting multiculturalism, volunteer activities	2009

(Continued)

Table 1.2 (Continued)

Region	Organisation Name	Ethnicity	Main Activities	Established
Sendai, Miyagi	PhilCom. Miyagi	Filipino	Safety confirmation, cultural exchange, etc.	2011
	Miyagi Prefecture Overseas Chinese "Same Boat" Club	Chinese	Safety confirmation, disaster information, cultural exchange between Japanese and Chinese	August 2012
	Miyagi Chinese Women Community Group	Chinese	Study support, cultural exchange between Japanese and Chinese	October 2016
	Chingudeul	Korean	offering language classes (partnered with Mindan Miyagi's Hangul School in 2014)	November 2010
Fukushima, Fukushima	Hawak Kamay Fukushima	Filipino	Cultural exchange, local volunteering	April 2011
Sukagawa, Fukushima	Tsubasa Support Group for Half Japanese-Chinese	Chinese	Language classes, parenting information, multicultural exchange	2011
Iwaki, Fukushima	Fukushima "Kokoro no Hashi" Multicultural Association	Chinese	Language classes, promoting multiculturalism, multicultural exchange	January 2014

I too was told by everyone [in my community] at first: 'Didn't you go home since you're a foreigner?' 'You're not one of the community members who live here'. I heard a lot of people telling me things like that. If I give up and go back [to South Korea], everything I have built up until now would be wasted. I tried hard and would like to live in harmony with everyone here, young and old.

Before the 3.11 disaster, Takako offered Japanese and foreign language classes in her local community funded by a grant from the Ministry of

Education, Culture, Sports, Science and Technology. After the 3.11 disaster, she was awarded other private grants to carry out various community recovery and regeneration projects involving local foreign nationals as well as Japanese people: visiting and checking in older adult victims/evacuees, starting a local community café and offering educational support for children in the disaster affected areas. She was also involved in the volunteer activities led by Korean companies. She played a key role. However, five years on, funding became difficult to secure, and she was unable to continue her NPO. After the community altruism started fading out, the public interest and support for her efforts to promote multiculturalism also gradually faded. She felt excluded and disempowered. Takako eventually left the area. Although we no longer had a chance to ask her the reasons why she left, it possibly was because her efforts to create a space to embrace multiculturalism in her community, help other migrants settle in and support local residents, in general, could not thrive in the long run.

Conclusion

Ten years on since the 3.11 disaster, it is no exaggeration to say that migrant women's empowerment, supported by disaster *communitas*, was not sustainable (Uekusa & Lee, 2020b). Their community regeneration and multicultural promotion efforts were well supported at the beginning by local communities and wider public, yet, over time, sustainability became dependent on resources and wider community dynamics. For long-term success, it seemed important for the community groups and, especially, the leaders to have a certain level of educational backgrounds and Japanese-language proficiency as well as their family support, financial resources and social capital.

Unfortunately, Jasmine from HKF also left Fukushima. Her reason for leaving was a fear of radiation and child wellbeing: cesium was detected in her daughter's urine. Her husband had stopped paying for her living expenses, telling her, '[Y]ou must be making money since you're so active outside the house!' Jasmine was indeed busy spending a lot of time outside the house as a leader of her community group, HKF, which supported others in need. Relying on a group she connected with during the 3.11 disaster, she moved to Osaka.

We do not criticise Takako's and Jasmine's decisions. This is not the goal of this chapter. Rather, this chapter has outlined how these decisions were likely shaped by external social factors that were beyond their control. Despite how hard these women worked, the long-term success of their community groups was limited. We highlight the fact that they both needed to leave their communities. Many rural ageing communities lost active members who had worked hard towards community regeneration, multiculturalism and migrant human rights. As emphasised throughout this chapter, despite the significant efforts made by these marriage-migrant women and external groups, the success and sustainability of their efforts tended to depend on larger social forces. In other words, if we do not understand these women in their social and cultural contexts and consider community, gender and racial dynamics, they could continue to be 'responsibilised' for their own fate and empowerment. Ten years on, we must again emphasise the need for disaster, migration and feminist scholars to take a critical approach to re-examine what has helped and hindered the empowerment of migrant women.

References

Akasaka, N. (2013). Imagining Japan in 50 years, from Tohoku (translation ours). *The Future Times*. Available at: http://www.thefuturetimes.jp/archive/no05/akasaka/ [Accessed 25 August 2021]

Aldrich, D.P. (2012). *Building resilience: Social capital in post-disaster recovery*. Chicago: The University of Chicago Press.

Ando, J. (2009). Migrant Spouses and Regional Communities in Rural Japan: A Case Study of Tozawa-mura, Yamagata Prefecture (translation ours). *GEMC Journal* (1), 26–41.

Arlikatti, S., Taibah, H.A. & Andrew, S.A. (2014). How do you warn them if they speak only Spanish? Challenges for organizations in communicating risk to Colonias residents in Texas. *Disaster Prevention and Management*, 23(5): 533–550.

Brighenti, A. (2007). Visibility: A category for social sciences. *Current Sociology*, 55(3): 323–342.

Cheng, C.M.C. & Choo, H.Y. (2015). Women's migration for domestic work and cross-border marriage in East and Southeast Asia: Reproducing domesticity Contesting Citizenship. *Sociology Compass*, 9(8): 654–667.

Constable, N. (2009). The commodification of intimacy: Marriage, sex, and reproductive labor. *Annual Review of Anthropology*, 38: 49–64.

Constable, N. (2012). International marriage brokers, cross-border marriages and the U.S. anti-trafficking campaign. *Journal of Ethnic and Migration Studies*, 38(7): 1137–1154.

Cutter, S.L., Boruff, B.J. & Shirley, W.L. (2003). Social vulnerability to environmental hazards. *Social Science Quarterly*, 84 (2): 242–261.

Enarson, E. & Fordham, M. (2001). From women's need to women's right in disasters. *Environmental Hazards*, 3: 133–136.

Enarson, E., Fothergill, A. & Peek, L. (2007). Gender and disaster: Foundations and directions. In Rodríguez, H., Quarantelli, E. & Dynes, R.R. (eds.). *Handbook of Disaster Research* (pp. 130–146). New York: Springer.

Faier, L. (2008). Runaway stories: The underground micromovements of filipina 'Oyomensan' in Rural Japan. *Cultural Anthropology*, 23(4): 630–659.

Hoschchild, A.R. (1989). *The Second Shift*. New York: Avon Books.

Ishinomaki City (n.d.). *Ishinomaki Plan for the Promotion of Multiculturalism [online]*. Available at: https://www.city.ishinomaki.lg.jp/cont/10053500/8296/goiken.pdf [Accessed 23 August, 2021]

JAR. (2011). Interim report: Employment support for foreign national women – Preparing for the nursing care certificate exam (translation ours) [online]. *Japan Association for Refugees*. Available at: https://www.refugee.or.jp/report/activity/2011/10/201110_-_2/ [Accessed 25 August 2021]

Kamoto, I. (2014). *Marriage migrants and 'multicultural coexistence' from the perspective of extraordinary events: Great earthquakes and divorces*. Kyoto: Kyoto Women's University. Available at: http://hdl.handle.net/11173/1631 [Accessed 25 August 2021]

Le Bail, H. (2017). Cross-border marriages as a side door for paid and unpaid migrant workers: The case of marriage migration between China and Japan. *Critical Asian Studies*, 49(2): 226–243.

Lee, F., Yamori, K. & Miyamoto, T. (2015). The relationship between local residents and media during recovery: Lessons from 'star disaster-affected areas' in Taiwan. *Journal of Natural Disaster Science*, 36(1): 1–11.

Lee, S. (2012a). The earthquake experiences and new challenges of 'multicultural families': How to understand the transnationality of marriage migrant women (translation ours). In Komai, H. & Suzuki, E. (ed.). *Immigrants and the Great East Japan Earthquake* (translation ours). Tokyo: Akashi Shoten.

Lee, S. (2012b). Gender and multiculturalism: Issues of marriage-migrant women in rural Tohoku (translation ours). *Tohoku University COE, GEMC Journal*, 7: 88–103.

Lee, S. (2013). Marriage-migrant women who invisibilize themselves (translation ours). In Hagiwara, K., Minagawa, M. & Osawa, M. (eds). *Taking Back Our Recovery: Women in Tohoku*. Tokyo: Iwanami Shoten.

Lee, S. (2018). Marriage-migrant Women and 'multicultural coexistence' in Tohoku: Overcoming 'othering' and 'invisibilizing' (translation ours). *The Bulletin of the Tohoku Culture Research Room*, 59: 73–87

Leong, K.J., Airriess, C., Chen, A.C., Keith, V., Li, W., Wang, Y. & Adams K. (2007) From invisibility to hypervisibility: The complexity of race, survival, and resiliency for the Vietnamese-American Community in Eastern New Orleans. In Bates, K.A. & Swan, R.S. (eds.) *Through the Eyes of Katrina: Social Justice in the United States* (pp. 171–188). Durham: Carolina Academic Press.

MacDonald, J. & MacDonald, L.D. (1964). Chain migration: Ethnic neighborhood formation and social networks. *The Milbank Memorial Fund Quarterly*, 42(1): 82–97.

Mainichi Shinbun. (2017, May 15). Filipinos introducing their culture [online]. *Mainichi Shinbun*. Available at: https://mainichi.jp/articles/20170515/ddl/k07/040/069000c [Accessed 25 August 2021]

Matthewman, S. (2015). *Disasters, Risks and Revelation: Making Sense of Our Times*. Hampshire: Palgrave Macmillan.

Matthewman, S. & Uekusa, S. (2021). Theorizing disaster communitas. *Theory & Society*, 1–20 (online first). doi.10.1007/s11186-021-09442-4

Minagawa, R. (2012). Working with foreigners in the affected area through law consultation activities (translation ours). In Komai, H. & Suzuki, E. (eds.). *The Great Eastern Japan Earthquake and Immigrants* (pp. 209–221). Tokyo: Akashi Shoten.

Ministry of Justice. (2010). Statistics for the resident foreign nationals [online]. *Ministry of Justice, Japan*. Available at http://www.e-stat.go.jp/SG1/estat/List.do?lid=000001111139 [Accessed 25 August 2021]

NSKK (n.d.). 'Let's walk together' Project [online]. *NSKK*. http://www.nskk.org/walk/?page_id=359 [Accessed 25 August 2021]

Osawa, M. (2013). How Science Council Japan reacted to the Great East Japan Earthquake from a gender perspective. *Trends in the Sciences*, 18(10): 30–35.

Parkinson, D. (2019). Investigating the increase in domestic violence post disaster: An Australian case study. *Journal of Interpersonal Violence*, 32(11): 2333–2362.

Petri, C. (2009). Translating disaster [online]. *The American Prospect*. http://prospect.org/article/translating-disaster [Accessed 25 August, 2021]

Portes, A. (1998). Social capital: Its origins applications in modern sociology. *Annual Review of Sociology*, 24: 1–24.

Sankei Shinbun (2011, March 26). Scared of radiation: Permanent resident Chinese wives rushing home, leaving their children behind (translation ours) [online]. *Sankei Shinbun*. https://megalodon.jp/ref/2011-0326-0310-38/headlines.yahoo.co.jp/hl?a=20110326-00000512-san-soci [Accessed 25 August, 2021]

Satake, M., Lee, I., Lee, S., Li, Y., Kondo, A., Saihan, J. & Tsuda, Y. (2015). Support initiatives towards cross-cultural families in Northern and Central Japan. *Journal of Nagoya Gakuin University Social Sciences* 52(2): 211–236.

Shimizu, A. (1987). *Ie* and *dozoku*: Family and descent in Japan. *Current Anthropology*, 258(4): s85–s90.

Solnit, R. (2009). *A Paradise Built in Hell: The Extraordinary Communities That Arise in Disasters*. New York: Viking.

Statistics of Japan. (2019). Demographic Statistics [online]. *e-Stat portal for Japanese Government Statistics*. Available at: https://www.e-stat.go.jp/stat-search/files?page=1&layout=datalist&toukei=00450011&tstat=000001028897&cycle=7&tclass1=000001053058&tclass2=000001053061&tclass3=000001053069&result_back=1&tclass4val=0 [Accessed 25 August, 2021]

Support Centre for Foreign Nationals, Tohoku University (2012). Report of 'Foreign National Disaster Victim' Survey Results Ishinomaki City [online]. Available at: http://gaikikyo.jp/shinsai/_src/sc550/90CE8AAA92B28DB895F18D908AT97v.pdf [Accessed 23 August, 2021]

Support Project for Foreign National Disaster Victims (n.d.). Ishinomaki City and Kesennuma City Survey Response Summaries [online]. http://gaikikyo.jp/shinsai/_src/139/90ce8aaa81e8bc90e58fc092b28db88fw8cv955c.pdf [Accessed 23 August, 2021]

Tierney, K. (2014). *The Social Roots of Risk: Producing Disasters, Promoting Resilience*. Stanford: Stanford University Press.

Torigoe, C. (2019). Racialized labels for foreigners, immigrants and foreign workers in Japan: Construction of racial 'others' in Japanese immigration discourses. *Studies in English Language and Literature, Seinan Gakuin University*, 59(3): 155–179.

Tsuchida, K. (2016). Can the vulnerable be community human resources? In Hasegawa, K., Hobo, T. & Ozaki H. (eds). *Recovery or Extinction? Disaster Recovery at a Crossroad*. Tokyo: University of Tokyo Press.

Uekusa, S. (2019). Disaster linguicism: Linguistic minorities in disasters. *Language and Society*, 43(3): 353–375.

Uekusa, S, & Lee, S. (2020a). Strategic Invisibilization, hypervisibility and empowerment among marriage-migrant women in rural Japan. *Journal of Ethnic and Migration Studies*, 46(13): 2782–2799.

Uekusa, S. & Lee, S. (2020b). Sustainable empowerment following disaster: A case of marriage-migrant women in the 2011 Tohoku disaster. In Fallaci, E. (ed.). *Women: Opportunities and Challenges* (pp.113–146). New York: Nova Science Publishers.

Uekusa, S., Matthewman, S. & Lorenz, D. F. (2022). Conceptualizing disaster social capital: What is it, why it matters and how it can be enhanced. *Disasters*, 46(1): 56–79.

Urano, M. (2014). Disaster process and vulnerability under the condition of the Great East Japan Earthquake: From the viewpoint of risk perception. *Bulletin of the Graduate Division of Literature of Waseda University*, 59(1): 71–86.

Yoshihama, M., Tsuge, A & Yunomae, T. (2013). "Violence against women and children after the Great East Japan Disasters: Results from a case-finding survey". In Japan Women's Watch (ed.) *Violence Against Women and Girls in Japan* (pp. 1–24). Tokyo: Japan Women's Watch.

Yoshitomi, S. (2008). *Multi-Cultural Symbiotic Society and the Power of Foreigners' Community*. Tokyo: Gendai Jinbunsha.

Wisner, B., Blaikie, P., Cannon, T. & Davis, I. (2004). *At Risk: Natural Hazards, People's Vulnerability and Disasters*. London: Routledge.

Villegas, F.J. (2010). Strategic in/visibility and undocumented migrants. In Sefa Dei, G.J. & Simmons, M. (eds.) *Fanon & Education: Thinking Through Pedagogical Possibilities*. (pp. 147–170). New York: Peter Lang Publishing.

11 The Social Amplification of Stigma in the Media after the Fukushima Disaster

Mikihito Tanaka

Memory and the Stigma of Fukushima

A large part of our social understandings is formed by imagined realities that are shared and experienced through the media (Couldry and Hepp, 2017). Furthermore, the decisive juncture of these realities might be the memories of devastating disasters that are engraved on our daily experiences. In recent decades, the frequency of large-scale disasters has increased. The focus of this chapter is the Fukushima disaster that occurred in 2011, but it was written in the context of another global pandemic disaster, COVID-19. The milestones of people's memories, which accumulate every day, are shaped by the memories of disasters. Indeed, 'we live in an age increasingly defined by media and disasters' (Pantti et al. 2012, p. 1).

Moreover, disasters, especially technological disasters, are often accompanied by stigma (Flynn et al. 2001). In Japan, numerous national debates took place after the nuclear disaster in the Fukushima Daiichi power plant triggered by the 2011 Great East Japan Earthquake (Fujigaki 2015). These arguments were not free from the memories of stigma as accumulated in Japanese history – scarred with nuclear bombs in Hiroshima and Nagasaki or the tragedy of fishermen who were accidentally exposed to radiation at the hydrogen bomb experiment (Valaskivi et al. 2019). Furthermore, in the contemporary media environment, the memories evoked through the Fukushima disaster were fermented in the 'hybrid media system' (Chadwick 2017) – the entangled complex amalgam of the mass and online media. There are many things to learn from considering the new relationships of a disaster, the hybrid media, and stigma, such as how the memory of stigma is framed, constructed, and handed down toward the next age in the shape of media content.

In this chapter, the media-led social construction of stigma regarding reactions to the Fukushima disaster is scrutinized, through an analysis of the trajectories of representations of space and time, as related to the 3.11 disaster. The analysis is based on the assumption that the media is the field of narratives represented by journalists, and collective memories are constructed and preserved through media (Zelizer and Tenenboim-Weinblatt, 2014). Furthermore, this chapter advocates the point of view that in a 'disruptive

DOI: 10.4324/9781003182665-14

media event' (Katz & Liebes 2007), like the Fukushima disaster, society preserves the painful memory and stigma throughout the production and consumption of disaster information (Pidgeon et al. 2003; Valaskivi et al. 2019).

Throughout the abovementioned process, the memories of Japanese people and the media were preserved and overwritten to the memory of nuclear weapons, which engraved stigma into the name 'Fukushima'. Two cases have been chosen in this chapter to observe the spatial and chronological media functions of constructing this stigma.

The first case considered is the 'maps' that were published in the newspapers after the 3.11 disaster. These maps represented scientific measurements and simulations of the radioactive contamination after the disaster and political countermeasures such as evacuation orders. The maps attached to the articles not only were additions but also framed the articles.

The second case concerns the side effects of the rejection of stigmatization by the people, which also gave rise to a discourse with a similar effect. While Japan was afraid of being stigmatized by other countries, as happened with Hiroshima or Nagasaki, the heroic tale of the 'Fukushima 50' became popular abroad, making its way into Japan sometime later. At first, the Japanese people and the media were embarrassed by the epic story of the Fukushima 50. However, they later abandoned the different narratives and adopted the Fukushima 50 epic, a simple myth that does not accept criticism.

This chapter explores the new relationship between a disaster, the (print) media and stigma. It presents an argument for how disaster narratives should be framed or how the controversial discourse should *not* be purified for representing disaster.

The Power of a Map: Arranged Stigma

Maps have power. From *Treasure Island* to *The Lord of the Rings*, maps have garnished incredible stories that liberated rather than limited the readers' imagination. The reader refers to the map to flesh out the story and imagine the fictional world's expanse beyond the written story. Nevertheless, the map in the real world also has the power to not only help our imagination but to also restrain our thinking. Maps simultaneously miniaturize, flatten, and generalize a space (Robinson et al. 1995, p. 5). However, a map is not limited only to being a diminutive copy of the world; under the consensus that it corresponds to reality, a map also creates reality. Whenever we look at a map of a land we have never visited, we would not doubt that the actual scenery is precisely the same as detailed on the map. In this way, maps are socially produced and socially functioning texts and have an actual effect on storytelling and the social construction of reality. Maps confine our imagination to the story being constructed.

Alongside the Fukushima disaster, the socio-technical representation of the resultant 'radioactive contamination' increased in Japanese media (Tanaka,

2019). Maps are strongly connected with the memory of hazards. The maps published in the Japanese newspapers after the Fukushima disaster textualized a spatial memory, involved ideology of scientism based on numbers, and limited the range of the story. It worked as a political frame to restrict media narratives. In this section, the spatial effect of the social functioning text is examined through the following trajectories of the maps published in Japanese newspapers after the Fukushima disaster, that is, a journey in which Japanese society *othered* Fukushima and allowed it to be stigmatized.

The Map and the Fukushima Disaster

One of the most famous maps related to the Fukushima disaster among the Japanese is the 'Radioactive Contamination Map' made by Yukio Hayakawa, a geology professor at Gunma University. The map visualized data from the Ministry of Education and first appeared in April 2011. This free downloadable map was revised eight times in the following two years and widely distributed in Japanese society after its first publication. However, beyond Hayakawa's intent to dedicate his expertise to the public (based on the interview with Professor Yukio Hayakawa with Louise Elstow, 2018), soon the map was politicised: it was given ideological meanings by frequent quotes in the media by anti-nuclear activists. Nevertheless, for a person who does not share such a context, this is merely a map showing Japan's soil contamination pattern.

This contextualization of Hayakawa's map indicates that maps can be political actors and mediate stigma through their function of fixing the invisible menace into visual hazard information. However, how are discursive effects weaved in the maps and affect the disaster's media framing? In this section, the effect of the map is investigated through a content analysis of the maps embedded in newspaper articles after the Fukushima disaster. This is an empirical study of how the maps framed the imaginary narratives of the hazard in the minds of both the journalists and the public.

Two national newspapers were chosen to investigate this process. The *Yomiuri Shimbun* (hereafter the *Yomiuri*) is the most popular newspaper in Japan. It is a centre-right newspaper with a circulation of approximately 8 million per day. The *Asahi-Shimbun* (hereafter the *Asahi*), a liberal newspaper, is the second-largest newspaper in Japan with a circulation of approximately 6 million per day (numbers based on the report from Japan Audio Bureau of Circulations, March 2021). The articles from four years after the disaster (from March 11, 2011, to March 31, 2015), including maps of radioactive contamination, were gathered from the newspaper database. After data cleaning, 1,545 articles (including 611 articles from the *Yomiuri* and 934 from the *Asahi*) were extracted from the initially collected 4,000 articles. These accumulated articles with maps were classified into six categories (Figure 11.1) In the next section, we follow the trajectories of the maps after the disaster.

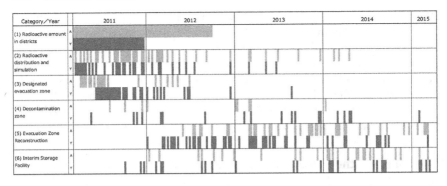

Figure 11.1 The advent pattern of the maps related to the Fukushima disaster in each newspaper.

The Trajectories of the Radioactive Contamination Maps

The maps of radioactive contamination, as presented within these newspapers, are strongly linked to the memory of the Fukushima disaster among Japanese people. However, the maps are not consistent – they represent different reality and privilege different aspects of disaster in their representations – and as such can be divided into at least six categories. In the first few weeks, these were '(1) radioactive amount in districts' maps, which included many dispersed points with numbers of measured radioactive dose rates on the map. Immediately after the disaster, maps of this type published in the *Yomiuri* and the *Asahi* covered a vast area of eastern Japan, from Yamagata prefecture in the north to Tokyo and Yokosuka in the south. However, these *points* were, over time, replaced with artificial geometrical *planes* on the map. To show the estimated tendency of radioactive spreading, concentric circles were drawn from the epicentre, the Fukushima Daiichi nuclear power plant, with radii from 2 to 80 km (50 miles). These circles were based on political estimation rather than scientific evidence, in addition to being based on the Japanese government and the US Army's prepared evacuation scheme.

Nevertheless, these maps gave a more 'scientific' impression (i.e. imparted a sense of objectivity) based on their geometrical expression. However, these geometric and political definitions of the hazard area were soon *scientifically* legitimized and substituted by the results of radioactive field measurement and computer simulations: there were '(2) radioactive distribution and simulation' maps. The supercomputer-based 'SPEEDI (System for Prediction of Environmental Emergency Dose Information)' maps were the precursor of this change (first appearing on March 24, 2011, in both the *Yomiuri* and the *Asahi*). The predicted tendency of radioactive spreading based on a simulation of SPEEDI merged with the concentric circles in the maps above. At this point, the data were chosen from the different simulation results, but later scientifically legitimized with the actual radioactive measurement of soils (Sugawara and Juraku, 2018; Sugawara and Juraku 2021). Namely, the contamination maps supposedly represented 'scientific data', but in reality, they were amalgams of political and scientific decisions. However, this category of maps decreased in 2012 and was never mentioned after 2014.

Later, the maps covered rather political issues and scientific evidence was contributed to legitimate the decision. The first of this type was '(3) designated evacuation zone' maps. As the name indicates, the maps in this category reflected the political decisions regarding the evacuation plan. However, aesthetically, this type was a successor of the '(2) radioactive distribution and simulation' maps. Based on the concentric circle maps at the early stage, the designated evacuation area maps defined the 'Evacuation Order Zone (20 km radius around the Fukushima Daiichi nuclear power plant)' and 'Planned Evacuation Zone (30 km radius, same as above)' and visualized the municipal orders. After the advent of these maps, the following types of maps legitimized the political decision: '(4) decontamination zone' maps that reflected the plans and progress of the decontamination of soils and the subsequent '(5) evacuation zone reconstruction' maps. Throughout this process, the area names and their borders in the damaged area were shown in bold lines. Namely, through the divergence of these political and scientific estimates, the *objective* map of contamination was defined. Finally, '(6) interim storage facility' maps, the ex post facto maps, became salient. These maps were called interim but were dedicated to establishing the scenery of the enormous numbers of piled huge plastic bags containing contaminated soil in Fukushima.

The most significant change that could be read from these trajectories of maps is that the maps minimized the whole picture of the Fukushima disaster. The disaster was first depicted as a national-level hazard that occurred in the whole of East Japan but was later curtailed into a problem relating to local Fukushima, and even later, limited only to the coastal area of Fukushima. The overall changes were not only *magnifying* processes in cartography but also *miniaturizing* the hazardous area. The narratives of the maps concentrated and diminished the nation-level hazard for most Japanese people, shutting the related problems and stigma into a very small area (Figure 11.2).

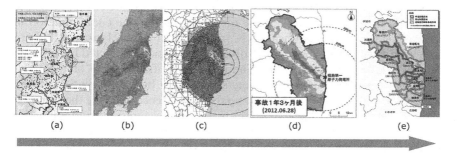

Figure 11.2 Trajectories of the 'Contamination Map'.

The Framing Effect of Maps

Maps are not merely secondary of articles, but they also reflexively frame articles and newspapers per se. The *objective* maps first framed the Fukushima disaster as a national issue, but the disaster was gradually being reframed into Fukushima's problem alone throughout a magnifying process. The contextualization of the articles and maps also reflected this tendency. For example, the page on which an article appears reflects the value of information in each newspaper. In both the *Yomiuri* and the *Asahi*, the articles containing maps were initially mentioned in the 'national issue' pages, which gradually changed to 'individual, societal issues' pages within a year and a half after the disaster. This change reflected the change in the framing of the national disaster that was gradually altered into the local disaster.

A more explicit variety of news values was observed. Whenever the articles with maps that included Tokyo or Kanagawa were carried on the 'national issues' page, the maps were printed in colour as these cities constituted the primary *market* of these national newspapers. However, these were gradually monochromatized along with the reframing of the disaster, which degraded it from a national-level issue to Fukushima's issue.

These changes are the consequence of news value after various information construction processes. The news is not only constructed inside the newsroom but also the accumulated interests of society affect complicated kinetics of news production. The state of framing and the public interest have a reflexive relationship with each other. For example, journalists who have been pursuing the Fukushima disaster for a decade share a common complaint: the more time passed, the more public interest in the disaster dwindled (Oketa 2021). National media vigorously reported the disaster in the first few years after 2011, but reports rapidly decreased (Shineha and Tanaka 2014; Tanaka 2014, 2019). Moreover, before this change, there was an apparent decline in the popularity of articles about the disaster on the internet and the interest of public opinion polls (Tanaka et al. 2012; Yamakoshi 2017). The phenomenon of the othering of Fukushima's issue by most Japanese is recursively based on the journalistic contents' unpopularity.

Additionally, there is another axiomatic function of the compound effect of the state of framing and the public interest that focuses on the maps. It can be seen that the trajectory of the maps shown in the above section (also see Figure 11.2) followed the scientific understanding of the contamination in the early stage and the political decision in the later stage. However, it is hard to ignore the fact that the myth of objectivity of science affected the framing of the Fukushima disaster throughout the scientific mapping of the early-stage contamination. Modern society is constraint by the convention of first measuring with science and then using that as evidence to define the situation (Porter, 1996). It was inevitable that the *point* measurement of radioactive dose rates was followed by the scrutinized measurement and visualized as a gradient of contamination as *plane figures* on the map. Nevertheless, it was also an eloquent fixing process to scientifically legitimize and stigmatize the hazard to the actual region symbolized on the map.

Hereafter, science gradually became merely a tool to legitimize the political othering of Fukushima. On April 12, 2012, the Japanese government declared that the acceptable radiation level in the designated evacuation area was 20 mSv/y. The *Yomiuri* represented the areas with the radiation level of 20 mSv/y on the map in bold lines with shades. However, there was no clear definition between the safe and the danger zones in science. This 20-mSv/y standard was politically defined to reduce the fatal radiation risk of death until 65 years of age to lower than a person per 1,000 (Murakami et al., 2014). The carefully designed standards by the political and scientific elites, the consequence of the scientific measurement, were applied to the real world, mediated by the flat-shaded area on the map.

Resisting the Framing of Maps

Hitherto, we followed the trajectories of maps in the *Yomiuri* and the *Asahi* to investigate the objective maps as a miniature of the actual world that had the ideological effect of fixing the stigma. However, it is also known that the ideological stance of news media can also affect their content. The *Yomiuri* and the *Asahi* are known for their dichotomous stance in Japanese media as conservative and liberal, respectively (Yamada, 2013; Nanasawa, 2016). The newspapers have historically covered nuclear power in a contrasting manner: the *Yomiuri* promoted nuclear power from its inchoate introduction to Japan in the 1960s. The *Asahi* maintained a suspicious stance towards nuclear power for many years. (However, it must be noted that the *Asahi* slightly changed its stance since the end of the last century, adopting the 'Yes, But' stance [Jomaru, 2012], in which they admitted the advantage of the nuclear power for tackling global warming but continued being wary regarding its safety.) Therefore, it should be noted here that such differences in political ideology manifested in different ways when utilizing maps in the news coverage of the Fukushima disaster.

However, the results of the analysis indicated that, in respect to the nuclear aspects of the Fukushima disaster, there are only slight differences in the news coverage of these two newspapers. For example, the *Yomiuri* stopped reporting the '(1) radioactive amount in districts' maps half a year after the disaster, while the *Asahi* continued publishing this type of map a year longer than the *Yomiuri*. Nevertheless, beyond the expected level, no significant ideological perspective of each media was clearly distinguished regarding the map usage of each media. Additionally, media policy is not only creating narrative styles. In reality, it can be observed that each newspaper also tried to resist the power of maps. To summarize this section, two cases are introduced as follows.

The first is the reframing of the scientific data on maps. During the post-disaster phase, the Japanese mainstream print media drew figures based on the data and maps provided by the Ministry of Education, Culture, Sports, Science and Technology (MEXT). This led to policymakers creating boundaries between safe and dangerous areas. Six months after the Fukushima disaster, both the *Yomiuri* and the *Asahi* published the map showing accumulated radio-active contamination in East Japan (September 30, 2011, and October 7, 2011, respectively). However, despite the original map from the MEXT showing the contaminated area of 3 million Bq/m^2 in red and the *Yomiuri* copying the map, the map published by the *Asahi* adopted a much lower threshold, 0.6 million Bq/m^2 area in red. Although all the maps were based on the same data, the high-risk zone was relatively smaller in the maps of the MEXT and the *Yomiuri* and more extensive in that of the *Asahi*. It is easy to criticize the *Asahi* by considering this difference as the ideological editing of the data. Additionally, it could be an example of manipulating the definition of risk to affect the impression of the maps from a different perspective. However, such an attempt to reframe the map by redefining the data should be maintained when utilizing maps in journalistic narratives. The maps have a strong power of framing in their ideology behind their pretence of objectivity.

The second is counteracting the power of maps by introducing victims' narratives. In this case, journalists tried to neutralize the framing of maps and expanded the meaning of maps with public discourse. For example, the *Yomiuri* delineated a story of a family whose members were experiencing difficulties because their house was located just outside the designated evacuation area (the *Yomiuri*, June 3, 2011). The family could neither receive reparation nor restart farming their land, but their neighbours, such as the shopkeepers in the designated evacuation area, were already evacuated, which made it hard for the family to survive. Furthermore, after the settlement of the 'special evacuation recommendation', a young family was evicted from their house. However, their old parents, who lived next door, could not move because their house was located beyond the border drawn on the map (the *Yomiuri*, August 4, 2011). The *Asahi* also reported similar cases, but it must be noted that even the *Yomiuri* depicted the power of maps affecting citizens' lives in addition to their appeasing stance toward nuclear power.

The preceding two cases may be only modest resistance to mitigate stigma. Indeed, maps have a strong framing power. The power of visualized maps naturally defines the range of hazards and the accompanying stigmatization by the journalists and their audiences. Simultaneously, with its visualization process, it also conceals assumptions and the resulting consequences: scientific reasoning of the depiction or the actual impacts of the disaster on the public became invisible to the audiences under the map's socially constructed objectivity. We cannot stop using maps to understand the world. Nevertheless, we must keep reinterpreting the maps and be cautious about the complexity shrouded behind the surface.

The Rise of the 'Fukushima 50' Myth as Counter-Stigma

The maps spatially brand stigma as a sketch of the real world. However, even the collective avoidance of stigma could chronologically give rise to an excessive myth. Consequently, this anti-stigma would paradoxically work to fix the stigma.

In March 2020, under the rising threat of the COVID-19 pandemic, a movie titled *Fukushima 50* was released in Japan. The movie was named *Fukushima 50* by the overseas media, referring to the 50 anonymous people who remained employed at the Fukushima Daiichi Nuclear Power Plant in crisis. The story focuses on Masao Yoshida, the then chief of the Fukushima Daiichi Nuclear Power Plant during the disaster and his colleagues, representing them as 'heroes who saved Japan'. Ryusho Kadota, the novelist who wrote the original novel on which the movie is based, said, '[W]ithout my book, I was worried that the workers dedication [to their duty] in the power plant would remain unknown. Japanese mass media mocked the disaster but never reported what happened inside the power plant' (*Fukushima Minyu Shimbun*, January 4, 2020, translated by the author).

However, it must be noted that the name 'Fukushima 50' *given by the foreign media* was exaggerated in this context. The other catchline of the movie was '[T]he hidden truth is revealed in the movie'. The assumption behind this discourse is evident: that the 'domestic media is not telling the truth'. However, the title and synopsis of the movie were attractive to the Japanese market because the story of the Fukushima 50 was already a familiar epic story popular among the public. The contradictory heroism that the 'anonymous people saved Japan' yet 'the truth is unknown' was widely shared among Japanese society.

Such stories could be necessary to dissipate the direct stigma passively attributed to Fukushima from other areas of Japan or overseas – the people struggling for the reconstruction of the radioactively damaged area are keen to escape the cruel stares, similar to the Japanese people who succeeded the memory of Hiroshima or Nagasaki (Valaskivi et al. 2019, pp. 61–79). Nevertheless, how was this epic myth constructed and shared? This section outlines the structuring process of the myth of the Fukushima 50, where and how it was born in cyberspace and amplified in the global media network, and how it was imported into Japan. This is an example of how disaster narratives can be spliced into public memory and how ideological arguments are presented through the contemporary media environment. This story shares a disgraceful event under a banner of *heroism* to people who struggled during the crisis.

The Advent of the Fukushima 50 Myth and Its Recurrence

First, it is necessary to provide a synopsis of the actual media event in which the discourse of the 'Fukushima 50' was constructed in hybrid media systems and later turned into a movie must be delineated.

The beginning of the case was March 15, 2011, the 22nd emergency press conference after the Great East Japan Earthquake, held at the Prime Minister's Office. In this press conference, the then chief cabinet secretary Yukio Edano declared that 800 workers temporarily retreated from the Fukushima Daiichi nuclear power plant, and '50 people are remaining' (*Asahi*, 15 March 2011). These words were immediately translated into

English and dispersed throughout social media. A few hours after this conference, cyberspace was filled with messages of cheers and laudations, followed by the posting of large quantities of illustrations supporting the Fukushima 50.

The news that 50 workers were staying at the most dangerous place on Earth and that the whole world was paying attention was also mentioned on US national television networks, such as CBS, and mutual interaction via social media was promoted. Moreover, on the following day, March 16, 2011, the newspaper the *New Zealand Herald* first introduced the term *Fukushima 50* in an article titled 'Heroes of Fukushima: 50 remain at Daiichi' (*New Zealand Herald*, 2011). On the same day, the *New York Times* (hereafter the *NYT*) described the workers with the title 'Faceless 50' (*NYT*, March 16, 2011), and this news disseminated throughout the world. Moreover, on the next day, March 17, 2011, seven Chinese newspapers quoted the *NYT* and dramatically reported about the story of 'the 50 braves in Fukushima'. Furthermore, on March 18, 2011, Japanese newspapers reimported these discourses from the world media and social media, with a direct translation of 'Fukushima 50'.

Three days after the chief cabinet secretary's first words, the discourse of the Fukushima 50 was produced in social media and amplified through mass media such as television networks and newspapers and recurrent to Japan. Moreover, throughout the circulation by way of translations across English, Chinese, and other languages, the context of the discourse was elaborated through dramatic rhetoric. However, the discourse soon disappeared from the Japanese media environment until the end of March 2011. As described later, it took three years for the discourse of the 'Fukushima 50' to reappear in the Japanese media.

The Rise of the Epic Tale

In more detail, the course from the emergence of the myth of the Fukushima 50 to its recurrence is filled with dramatic rhetoric. The sentences initially described by the chief cabinet secretary, Yukio Edano, in the conference were only the fact that 800 workers had retreated from the power plant and that 50 had remained at the plant. However, the discourse was reframed rhetorically under the banal epic story handed down from the past. In social media, the primary incubation media sphere, people throughout the world thought fondly about the anonymous heroes and modified the story in their own ways. Edano's utterances were incomplete but evocative, and there was a blank slate for individuals to enrich the story with their imagination. The vacant space was filled with eloquent expressions such as the 'nameless 50' or 'the last soldiers saving Japan', which were later handed over to the social media users.

Nevertheless, motifs adopted in the rhetoric were based on historical sentimental imaginaries. Moreover, culturally stereotypical similes, such as *samurai* or *kamikaze*, were also evoked and analogized in the public's imagination. Furthermore, the observers in the overseas social media sensed that there was an assumed image shared among the public. One of them was the tragedy of the firefighters in Chernobyl, who suffered fatal exposure while performing their duty. The other story shared was that of the Hollywood movie *Armageddon* (1998), famous for its plot in which petroleum miners dedicated their lives to saving the Earth.

When connected by the hashtag #FukushimaFifty, social media users collectively built an image through mutual interaction. There is no doubt that these people truly worried about and cheered for the workers at the Fukushima Daiichi Nuclear Power Plant. Even so, simultaneously, it is also clear that people shared and expected a cliched story that they were familiar with through the news and the movies: the epic tale in which a few anonymous, altruistic laypeople sacrifice their lives to save the many. It is difficult to deny this shared cruel expectation. For example, an enormous number of consumer-generated illustrations were made for the Fukushima 50 on social media, but most of their representations were tinged with the rich motif of death – nuclear plant workers wearing kamikaze pilot suits or gas masks as an obvious skull metaphor. Indeed, people were acclaiming workers in Fukushima but were also expecting a repeat of the tragedy.

Establishment and Amplification of Epic Tables by Press

Even if the users shared some images of the Fukushima 50, stories generated on social media were fragmented. Affected by people's imaginary narratives online, journalists wove them to the end of fructified epic poetry. As the vanguard, the *New Zealand Herald* officially called the workers who remained in the Fukushima Daiichi Nuclear Power Plant the 'Fukushima 50' and delineated them as the already-known heroes. The *NYT* article on March 16, 2011, which was soon quoted by the world's press, is filled with such rhetoric (*NYT*, March 16, 2011). While it refrained from the story nourished in social media that anonymous samurais were desperately grappling with the fatal nuclear power plant, the US nuclear expert quoted in the article praised the bravery of the remaining 50 Fukushima workers. It was crystallized as a magnificent tale of altruism by high-impact journalists. Even the lack of information set the drama off: the author of the article could not get a comment from the Tokyo Electric Power Company (TEPCO) that manages the Fukushima power plant, but this fact was also used as a prop to adorn the anonymity of the heroes in the story.

The representation of brave heroes was succeeded by later reports and these reports were even more embellished. Such furbelows reached a crescendo in the Chinese media. Throughout the news production of the Chinese press, all the vivid images introduced in prior reports were inherited, the seriousness of the situation was enhanced, and statements describing the experts' admiration for the self-sacrifice of the 50 workers were added. For example, appellations for the nuclear workers in Fukushima in the Chinese press were as follows: '50 braves of Fukushima' (translated by the author; originally '福島50勇士' in Chinese), '50 daredevils of Fukushima' (福島50壮士), '50 martyrs of Fukushima' (福島50烈士), and so on. Considering that tensions between China and Japan were rising before the disaster because of the territorial disputes in the East China Sea, the reactions were unusual. However, the escalation of these epithets reflects the main virtues of the Chinese cultural context: heroic bravery, fierce determination, and dedication of one's life to the public (Sheridan 1968; Denton 2017). These epithets were synonyms only given to heroes and familiar to Chinese people throughout their patriotic education.

Finally, an amalgam of the imaginary narratives revealing the cultural self-reflection of each nation state returned to Japan. The Japanese have a relatively strong interest in how they are perceived by people from other countries (Nakatani et al. 2003; Valaskivi 2016). The news that '[t]he Fukushima 50 acclaimed from the other countries' was welcomed by each media outlet and was introduced in Japan with a thrilling touch. However, an atmosphere of hesitation can also be observed in the tone of these articles. The situation at that time was too strained to innocently accept the epic representation of nuclear workers struggling at Fukushima. The English and Chinese articles mentioned earlier were reports transformed into epic tales, having only a little critical perspective about *why* the workers remained and *who* were these workers.

On the other hand, in Japan, just after the disaster, this laudation was far from reality; people felt embarrassed and had no composure to indulge in admirations from overseas. Other media coverage regarding this were reports of the detailed persona of the workers, the much-lauded Fukushima 50. Even the pro-nuclear conservative media quoted the voices of families worrying about their husbands or fathers. The *Asahi* compared the tale of the Fukushima 50 to Kenji Miyazawa's *The Life of Guskō Budori* (1932) – a story about an engineer who sacrificed himself to save people – and not only acclaimed its preciousness act but also desired that the consequence of the reality differs from the novel (*Asahi*, March 18, 2011). Such hesitancy toward the worship of heroes was not a ritualized phenomenon in the mass media. The song of praise moved people in social media; however, it was observed that repetitive expressions in limited numbers were used to pray for the safe return of the workers.

The representation of heroes was rapidly, collectively, and simultaneously constructed in cyberspace, accumulated and raised to the epic tale in the newspaper, and recurrent within Japan. However, in the crisis stage just after the accident, the heroic tale of the Fukushima 50 was not an 'unknown' story as the conservative critics stated in the campaign of the movie which instead embarrassed the then Japanese audiences as described later. Since then, as the Fukushima nuclear power plant gradually recovered from the crisis, the epic tale of the Fukushima 50 was gradually forgotten by the Japanese media until it gained new life in 2014.

The Enhancement after Defamation: The 'Yoshida Testimony' Case

When the Fukushima 50 was resurrected by the Japanese media in 2014, the sense of awkwardness toward this discourse was forgotten. Instead, throughout the argument described in the following, it is observed that the tale of the Fukushima 50 was perfected as a divine *myth*.

The beginning of this resurrection was a scoop by the *Asahi*, the Yoshida Testimony campaign. On May 20, 2014, the *Asahi* reported that they exclusively acquired a concealed interview record relating to the Fukushima disaster. The log was an audio and text record of Masao Yoshida, interviewed by the governmental Investigation Committee on the Accident at the Fukushima Nuclear Power Stations on July 22, 2011. Yoshida died of cancer in 2013, so his testimony about what happened at the site after the disaster was significant. The article's title was 'Yoshida Testimony: The Fukushima nuclear accident as told by plant manager Masao Yoshida'. The *Asahi* deplored this serial media campaign both through the newspaper and 'Scrollytelling'-styled website (http://www.asahi.com/special/yoshida_report/en/).

However, besides the journalistic value of the interview with respect to its uncovering process, the contents of the report were not so exclusive as to call it a scoop. The substance of the contents was already known through the past interviews of Yoshida and his relatives as well as previous reports. Nevertheless, the aim of this campaign was evident from its title of the first chapter: 'Who is there to halt nuclear reactors?: Reality of the "Fukushima 50"'. The *Asahi* tried to critically reframe the atmosphere of admiring the Fukushima 50 by delineating the chaotic situation after the accident.

Nonetheless, beyond the expectations of the *Asahi*, this campaign provoked fierce rejection from many other Japanese media outlets. In particular, conservative newspapers such as the *Yomiuri* and the Sankei Shimbun conducted a long campaign refuting the Yoshida Testimony serial, and criticisms also raged on the internet. The refutation's main focus was an expression in the article that stated that because of the miscommunication between the plant manager Yoshida and the workers, many workers withdrew against an order from Yoshida. After the onslaught, Tadakazu Kimura, the then CEO of the *Asahi*, apologized, stating that '(the article) gave the impression that many TEPCO employees had abandoned the plant because it reported that they had "withdrawn against the order," which was based on an erroneous appraisal made in the course of reading and trying to understand the Yoshida testimony' (Kimura, *Asahi*, 12 Sept 2014), and Kimura later resigned from his post.

This Yoshida testimony case is now remembered as a 'false' report, or 'fabrication' by more conservative people in Japan. However, the report is still available online, both in Japanese and in English. The original expression which provoked a number of disputes is still shown, and without scoring it, the 'corrected' expression is written in an awkward red manner. Probably, the reader without any preconception would feel curious that these original expressions were fiercely censured as fabrication and would also be puzzled that the angry critics were satisfied with such minor corrections. While the focus of this section is not the legitimacy or the rightness or wrongness of this campaign, it is important to examine the ripples that the debate cast on the image of the Fukushima 50. Even from this perspective, the nucleus of the critical focus was not the expression in the Asahi's article. The following is an extract from the Yomiuri

> The most severe problem is that foreign countries recognized workers in Fukushima as people who 'withdrew against the order'. Until then, the workers who stayed in their place were worshiped as 'Fukushima 50' and reported as brave people. After the *Asahi Shimbun* reported 'violation of the order', the reputation of the Japanese as people who do not run from a crisis was defamed.
>
> (*Yomiuri*, August 31, 2014, translated by the author)

Indeed, soon after the Yoshida testimony reported by the *Asahi*, the foreign press that had applauded the Fukushima 50 in 2011, including the *NYT*, immediately reported the *Asahi*'s scoop in headlines such as 'Panicked Workers Fled Fukushima Plant in 2011 Despite Orders, Record Shows' (*NYT*, May 20, 2014). In the end, the conservative media problematized the defamation of the epic tale rather than the expression of the report. The reluctance that was once shared among Japanese society when the representation of the Fukushima 50 drifted in 2011 was dismissed. There was only a desire for the heroism of the Japanese to remain intact, and when the condemnations drew apology and correction from the *Asahi*, the *myth* of the Fukushima 50 was completed in Japan.

Ways to Mitigate Stigma within the Disaster Narrative

This chapter investigated stigma, as framed and fixed by compressing space into images, or as gradually constructed by withstanding prejudicial situations.

As we saw throughout the study of maps, visualisation is frequently used to curtail the complexity under risk circumstances such as a disaster. However, this could also give a chance to the anxious mass audience to stigmatize the disaster to some region and its residents (Flynn et al. 2001: pp. 10–12). In the first half of this chapter, the maps, usually deemed to be an objective reference, could potentially fix the stigma with their framing effect were observed. However, when changing the standpoint, maps can be used for visualizing the stigma. By deciphering the narratives consisting of maps or seeking narratives *not* retrieved from the map, senders and receivers of the news can avoid stigma by deconstructing the situation.

In the latter half of this chapter, the process by which the workers who fought to bring the Fukushima nuclear accident under control became represented as heroes who saved the world was traced. At first, Japanese society was embarrassed about the epic tale from abroad but later accepted the story as a myth to compensate for the anxiety to been stigmatized. The myth of the Fukushima 50 became not only an epic tale of those who saved Japan from the nuclear disaster but also a *curse* that distract from the need of Japanese society to change and learn important lessons (Samuels, 2013). There is a need to stop viewing that a simple evil element leads to a disaster – in the post-disaster arguments, Japanese liberals tend to criticize TEPCO, and conservatives criticize the prime minister, Naoto Kan (e.g., Yamakoshi 2017); otherwise, adherence to the mighty epic myth hinders the balanced, resilient society from learning lessons. A column from the *Mainichi-Shimbun* briefly summarized the perspectives regarding the mythologizing and the subsequent defamation of the Fukushima workers as follows:

> It has been successively reported that workers who reacted to the nuclear power plant accident who were called as 'Fukushima 50' heroes, are now being criticized for 'withdrawing against the order'. However, both perspectives lack sincerity. The most important lesson that we must learn from the Yoshida testimony is how to avoid the next accident. We do not need heroes in nuclear power plants.
>
> (*Mainichi Shimbun*, September 25, 2014, translated by the author)

Epic tales give us the power to withstand stigmatization. However, simultaneously, if the narrative is enshrined at the myth level, it would be disguised as a distant memory beyond reflection. To avoid stigmatization by the heroic myth, the narrative's mellow venom must also be remembered to learn important lessons.

Epilogue: Stigma in the Age of Hybrid Media

The preceding analyses were based on the content analysis of newspaper articles that are still vital information hubs in hybrid media systems. Nevertheless, it should not be forgotten that the social media environment that produces, reproduces, and sorts and stores narratives plays an essential role in the contemporary stigmatization process, as shown in these analyses. For example, every year, when March 11 approaches, many discourses appear to criticize the stigma given to Fukushima. However, the analysis of the actual public discourse tendencies in social media indicates that uncivil posts that actually stigmatize Fukushima are rarely observed. If anything, most posts related to stigma are against the many people who try to stigmatize Fukushima.

Here is a paradox similar to the Fukushima 50 case. The origin of the polarization of society is the distinction between *us* and *them*. The life span of a specific piece of information in social media is short, and collective memory can only remain consistent through repetitive narratives. Ironically, narratives of stigma are preserved by shunning posts for such incivility by the public.

After the internet penetration into society in this century, we are in the mediated turmoil that makes it hard to trust the mass media, and even though we are connected via ubiquitous online mediums, we experience othering through social media. In such circumstances, it is becoming more difficult to free oneself from the media effect of the framing of stigmatization through maps or from falsifying memory and seeking intact heroic myths, as was observed in the preceding case studies. The path to overcoming the challenge of embracing the pain of unstable memory without stigma lies amidst this turmoil. The only way to overcome the divided imaginaries would be derived from the accumulated struggle of reflexive efforts to grasp the memory and stigma of public opinion in turbulence.

References

Bruckheimer, J., G. A. Hurd, and M. Bay (Producer). (1998). *Armageddon* [Motion picture]. United States: Touchstone Pictures.

Chadwick, A. (2017). *The Hybrid Media System: Politics and Power*. London: Oxford University Press.

Couldry, N. and Hepp, A. (2017). *The Mediated Construction of Reality*. Cambridge: Polity.

Denton, K.A. (2017). 'Heroic Resistance and Victims of Atrocity: Negotiating the Memory of Japanese Imperialism in Chinese Museums', *The Asia-Pacific Journal*, 5(10), 2547.

Flynn, J., Slovic, P. and H. Kunreuther. (2001). *Risk, Media and Stigma: Understanding Public Challenges to Modern Science and Technology*, London: Earthscan Publications Ltd.

Fujigaki, Y. (ed.) (2015) *Lessons from Fukushima: Japanese Case Studies on Science, Technology and Society*, London: Springer.

Fukushima Minyu Shimbun (2020). 'Gensaku Kadota Masao san "Fukushima ga Nihon wo sukutta" [Masao Kadota says: Fukushima saved Japan]', 4 January. Available at: https://www.minyu-net.com/serial/fukushima50/FM20200104-447695. php [Japanese]

Japan Audio Bureau of Circulations (2021). '*Shimbun Tsukibetsu Report*' [Monthly Newspaper Circulation Report], March 2021.

Jomaru, Y. (2012). *Genpatsu to Media: Shimbun Journalism Nidome no Haiboku.* [*Nuclear Powerplant and the Media: The Second Defeat of the Newspaper Journalism.*] Tokyo: Asahi-Shimbun Shuppan. [Japanese]

Katz, E., and T. Liebes. (2007). "No More Peace!': How Disaster, Terror and War Have Upstaged Media Events'. *International Journal of Communication*, 1(1), 157–166.

Kimura, T. (2014) "I apologize to our readers and other people concerned By TADAKAZU KIMURA / President of The Asahi Shimbun," 12 September 2014. http://www.asahi.com/shimbun/20140912english.pdf (Retrieved February 4, 2021)

Miyazawa, K. (1932). *Guskō Budori no denki* [The Life of Guskō Budori]. Jido Bungaku, Vol. 2 [Japanese]

Murakami, M., T Nagai, K. Ono, and A. Kishimoto (2014). *Kijyunchi no karakuri* [Tricks of the Standard Numbers]. Tokyo: Kondansya. [Japanese]

Nakatani, T., H. Takahashi, and T. Kawakami. (2003). *National Identity ron no genzai [The locus of the National Identity Studies]*, Kyoto: Koyoshobo. [Japanese]

Nanasawa, K. (2016). *Televi to genpatsu houdou no 60 nen* [*60 years of the nuclear power plant reporting in Television*], Tokyo: Sairyusya. [Japanese]

Oketa, A. (2021). *Media ha 'saika' wo doukatarouto surunoka* [How do the media talks about disaster?] In R. Shineha (ed.) *Saika wo meguru 'kioku' to 'katari'* [*Memories and Narratives Surrounding the Disaster*], Kyoto: Nakanishiya. [Japanese]

Pantti, M., K. Wahl-Jorgensen, and S. Cottle. (2012). *Disasters and the Media*. Peter Lang.

Pidgeon, N. R. E. Kasperson, and P. Slovic. 2003. *The Social Amplification of Risk*. Cambridge University Press.

Porter, T. M. (1996). *Trust in Numbers: The Pursuit of Objectivity in Science and Public Life*. Princeton University Press.

Robinson, A. H., Morrison, J. L., Muehrcke, P. C., Kimerling, A. J., & Guptill, S. C. (1995). *Elements of Cartography* (6th Edition). London: Wiley.

Samuels, R. J. (2013). *3.11: Disaster and Change in Japan*. New York: Cornel University Press.

Sheridan, M. (1968). "The Emulation of Heroes." *The China Quarterly*, 33, 47–72.

Shineha, R. and M. Tanaka. (2014). "Mind the Gap: 3.11 and the Information Vulnerable," *The Asia-Pacific Journal*, 12(7), 4.

Sugawara, S. and K. Juraku. (2018). 'Post-Fukushima Controversy on SPEEDI System: Contested Imaginary of Real-Time Simulation Technology for Emergency Radiation Protection' in S. Amir (ed.) *The Sociotechnical Constitution of Resilience*, Palgrave Macmillan.

Sugawara, S. and K. Juraku. (2021). 'Structural Ignorance of Expertise in Nuclear Safety Controversies: Case Analysis of Post-Fukushima Japan', *Nuclear Technology*, 207, 1423–1441.

Tanaka, M. (2019). "Toward Geography of Mediated Affect: Discursive Spaces and Emotional Dynamics," In Valaskivi, K., A. Rantasila, M. Tanaka, and R. Kunelius. *Traces of Fukushima: Global Events, Networked Media and Circulating Emotions*. Palgrave Macmillan.

Tanaka, M. R. Shineha, K. Maruyama. (2012). *Saigai jakusya to joho jakusya* [*Vulnerables of the disaster and information*] Tokyo: Chikuma-shobo. [Japanese]

The Asahi Shimbun (2011). 'Genpatsu jiko no saigo no toride [The last stronghold of the nuclear disaster].' March 18, 2011. [Japanese]

The Asahi Shimbun, "Full Record of the press conference by the Cabinet Secretary Edano, 11 am, 15 Mar. 2011". n.d. http://www.asahi.com/special/10005/TKY201103150218. html (Retrieved December 24, 2020) [Japanese]

The Mainichi Shimbun. (2014). 'Science Café: Hero wa iranai' [Science Café: We don't need hero], 25 September. [Japanese]

The New York Times, Bradsher, K. and Tabuchi, H. (2011). 'Workers Brave Radiation Risk at Failing Japan Reactors', 16 March.

The New York Times, Fackler, M. (2014). 'Panicked Workers Fled Fukushima Plant in 2011 Despite Orders, Record Shows', 20 May.

The New Zealand Herald. (2011). 'Heroes of Fukushima – 50 remain at Daiichi', 16 March.

The Yomiuri Shimbun (2014). '*Tenken Yoshida Chosho: "Syoin tettai" sekai ni gokai* [*Examining Yoshida testimony: "Workers withdrawn" lead misunderstanding to the world*', 31 August. [Japanese]

Valaskivi, K. (2016). *Cool Nations: Media and the Social Imaginary of the Branded Country*. London: Routledge.

Valaskivi, K., A. Rantasila, M. Tanaka, and R. Kunelius. (2019). *Traces of Fukushima: Global Events, Networked Media and Circulating Emotions*. London: Palgrave Macmillan.

Yamada, K. (2013). *3.11 to media: Simbum, TV, Web wa naniwo dou tsutaetaka*. [*3.11 and Media: How and What newspapers, TVs and Web reported*], Tokyo: Transview. [Japanese]

Yamakoshi, S. (ed.) (2017). *Sengo Nihon no Media to Genshiryoku Mondai* [*The Post-war Japanese Media and Nuclear Problems*], Kyoto: Minerva Shobo. [Japanese]

Zelizer, B. and K. Tenenboim-Weinblatt. (ed.) (2014). *Journalism and Memory*, London: Palgrave Macmillan.

Part IV

Community Engagement and Wellbeing

12 Theatres of Resilience
Schoolchildren as Actors in Community Development in Fukushima

Alison Lloyd Williams and Aya Goto

Introduction

The 3.11 disaster led to a remapping of Fukushima as a place to live, work and play. For the generation of children growing up in the wake of the disaster, visible signs of its legacy and the slow recovery process would become an everyday feature of the landscape: from the thousands of suited decontamination workers who moved into the region to the fields full of bagged up toxic waste to the radiation monitors placed outside schools and in other public places. At the same time, children's bodies were seen as another post-disaster landscape to be worked on as part of the recovery effort, with their physical health closely monitored for signs of impact. Despite this heightened scrutiny, there has been less research into children's *own* perceptions of their lives in Fukushima and *their* ideas for community development as the area continues to recover and rebuild. Indeed, amid the climate of 'fear, anxiety and hope' felt by ordinary people following the disaster (Koikari, 2020: 4), children have become a 'concentrated site of resilience politics as they bear the nation's future', but dominant discourses have tended to obscure alternative viewpoints, including those of young people themselves (ibid: 95).

Definitions of *resilience* in disaster management are diverse, but we take from Mort et al.'s work in framing resilience as more than individual adaptiveness and as something achieved collectively and creatively: 'the fruit of empowering and creating interdependence, solidarity and agency, especially with those groups that are the most silenced and marginalised' (2020: 7). This approach emphasises the 'voices, perspectives and actions' of children as a marginalised group (ibid: 16), recognising the knowledge, skills and insights they can bring to discussions about how societies prepare for, respond to and recover from disasters. Research has shown for example, that children recognise that things do not simply 'go back to normal' after a disaster and that they understand that communities need to adapt to become more resilient (Mort et al., 2018: 434). Methodologically, there is an increasing turn to participatory work with children in disaster contexts in order to support them to explore and communicate their experience and ideas and engage with other stakeholders (Lloyd Williams et al., 2017). This is recognised as bringing benefits both to the children and wider society (Amri et al., 2018: 239).

DOI: 10.4324/9781003182665-16

These reflections underpinned the approach to our project which began in 2016, when our team of public health and social science researchers came together with local educational professionals to work with a class of Grade 6 students at a school in Date City, Fukushima. The aim was to use participatory theatre methods to discover and promote children's views about their community, five years after the disaster. Since this initial phase, the work has gone on to explore how teachers can embed these methods into schools as an approach to community resilience building in Fukushima. The team has also continued to develop the use of theatre methods with young people in response to other health crises, including the COVID-19 pandemic. The work has shown the potential of the approach in supporting young people to voice their experiences and contribute to resilience building – to become 'actors' in community development in Fukushima.

Children and Schools as Actors in Resilience Building in Fukushima

The effects of the 3.11 disaster on children have not been fully investigated. Most studies focus on children's physical health, in particular the potential impact of radiation (e.g. Nishikawa et al., 2020; Ohtsuru et al., 2019) and the increase in childhood obesity connected with a lack of outdoor exercise (Itagaki, 2017; Yamamura, 2016). Other research has focused on psychological impacts, such as the effect of bullying on children from affected regions (Lieber, 2017; Oe, 2019). These studies reflect a wider trend in work on children and disasters, as Wisner et al. discuss, in focusing on 'children's trauma and symptomology, rather than on investigating processes that enable children to respond adaptively within their social and environmental contexts' (2018: 6). Our team was interested to learn how children living in the shadow of the 3.11 disaster were adapting to life in a community in recovery, where changes brought about by the disaster would be part of 'normal' life for them. Our project aimed to fill the gap Wisner et al. identify by researching *with* children, rather than *on* them, and inviting them to explore their own perceptions of their communities. Further, we wanted to bring children's voices into conversations around recovery and resilience building by sharing their experiences and ideas with other stakeholders – teachers, parents and local leaders. In that way, we were interested to see how children could play an active role in 'community health' following the disaster (Reich & Goto, 2015).

Our work with the children started from the perspective that children have a right 'to participate in all matters that affect them' (OHCHR, 1989). This rights-based approach configures children as citizens or actors in disasters with their own knowledge and insights. Research with disaster-affected children has shown that this expertise can inform wider societal approaches to how we live with disaster risk (Mort et al., 2018), and there is increasing recognition of the role children can play in disaster risk management as 'risk communicators, community mobilisers, and agents of change' (Amri et al., 2018: 242–3). This view is underlined in the 2015 United Nations Office for Disaster Risk Reduction Sendai Framework, which states that '[c]hildren and youth are agents of change and should be given the space and modalities to contribute to disaster risk reduction, in accordance with legislation, national practice and educational curricula' (UNDRR, 2015).

Situating children as actors in disaster resilience requires approaches that create the 'space and modalities' for them to act, and there is therefore an increasing use of participatory methods to support children to express and share their experiences. For children, participating in disaster resilience work can promote critical thinking, problem-solving and social development (Pfefferbaum et al., 2018), forms of learning and skills development that sit well within either formal or non-formal educational settings. But this work can also support children to 'process' events and 'create a new shared narrative', taking the focus beyond the self to the 'wider community' (Gibbs et al., 2017: 149). In that way, involving children can potentially inform new ways of being and doing and help strengthen networks of disaster preparedness (Pfefferbaum et al., 2018; Wisner et al., 2018).

If children can play a part in disaster reconstruction, so schools can become local hubs for resilience building, first by providing active learning spaces for children to reflect on their experiences together but then also as focal points for sharing that learning as a stimulus to action with the wider community. This approach aligns with Percy Smith et al.'s concept of 'participatory citizenship education' (2019: 194) in which children develop as citizens through the practice of engaging with 'matters that affect them' (OHCHR, 1989) rather than learning 'about' citizenship through traditional formal learning. As the authors point out, this requires a different pedagogic approach from the traditional dynamic between adults/teachers and children/students: the role of the teacher shifts from that of the 'expert' to one of a facilitator, supporting the children to express and explore their own expertise in a way that 'repositions young people in society as diverse but equal citizens with meaningful contributions to make' (ibid: 195).

In putting forward this idea of schools as hubs for community resilience building, we build on work by Parmenter which argues that the 3.11 disaster reinforced the importance of schools as key 'citizenship centres' in Japanese society and children as 'active citizens', both within school and the wider community (2012: 11–12). Parmenter suggests that, while the notion of 'school as community' has always been important in Japan, the disaster added a vital new role – that of the 'school in the community' (2012: 13). She notes that in the days and weeks following the disaster, schools served as key hubs for the response and recovery effort, and children played an active part in that work. In doing so, the students drew on the sense instilled in all children at Japanese schools that they are active members of the school community with roles to play in the day-to-day running of that community. As schools broadened their remit to support the wider local community, the students stepped up their involvement according to the methods they were used to in school. It follows, therefore, that schools could build on that role by becoming focal points of community preparedness and resilience building with children as key players in that activity.

Participatory Theatre as a Way to 'Map' Children's Lives in Fukushima

There has been considerable research into the ways in which participatory mapping can be used with children as a way to understand their lived experiences and to see their communities from their perspective (see, e.g., Martz et al., 2020). Such work situates young people as 'already-knowledgeable subjects in their everyday lives' and explores how these processes 'can enhance our understanding of young people's social and material worlds' (Gordon et al., 2016: 558–9). This approach reflects a wider concern in the social sciences with the significance of space and place – seeing these not merely as 'the background to human action' but as shaping and being shaped by the people who move in and out of them (Djohari et al., 2018: 352).

Participatory mapping focuses increasingly on 'exploring how practices, identities and experiences emerge through, rather than apart from, the world we live in' (ibid), suggesting a more relational view of human agency and a more dynamic view of space as a 'process', always active and constantly changing (Bartos, 2013: 89). As Bartos notes, '[b]oth our sense of self and our sense of place are perpetually being felt, formed, dislodged, and (re)created. Therefore, attention to sensing place has the potential to help learn more about our sense of self and our sense of place' (2013: 91). Bartos suggests that children engage more through the senses and thus the present compared to adults, who rely more on intellectual engagement and memory. Thus, a methodology that invites children to map their community through their sensing body can be an effective way to capture an understanding of the world as they experience it.

Our project used participatory theatre, a practice concerned with the actions of the sensing body in space, as an innovative way for children to 'map' their community. The methodology drew specifically on Theatre for Development, an interventionist blend of practical drama activities and discussion, used to explore community issues and mobilise participants to social action (Lloyd Williams, 2015). Through this iterative, creative practice we invited the children to physically re-create and explore aspects of their community, using the building blocks of performance: their bodies and voices and the space around them. Participatory theatre has not often been applied in post-disaster settings, particularly in Japan. Where it has been used, such as O'Connor's work after the Christchurch earthquake in New Zealand, it's been in a more therapeutic way (O'Connor, 2015). However, rather than looking back on the disaster and focusing on individual recovery from trauma, our project invited children to explore together their community *now*, reflecting on their current lived experience and their ideas for future change.

Participatory theatre provides a safe space for children to re-create and explore the'world outside' through role-playing different people and objects, creating landscapes through movement and sound, showing relationships and creating different atmospheres. This work invites participants to think about space and their bodies in that space, using what Gordon et al. call 'critical spatial thinking' (2016: 559). The children step in and out of the drama, creating scenes or actions, discussing them and changing them, adding physical details, taking them away, moving them to different places and connecting or disconnecting them from other elements on stage.

This process of inventive play – what Geilhorn calls the 'utopian possibilities' of theatre (2017: 162) – also allows children to circumvent the boundaries of reality and fiction, to play with the 'real' spaces they are exploring, rehearse different ways of doing things and map new, imagined landscapes. This 'liminal space for imagining alternative realities' allows for an indirect exploration of social issues, which Geilhorn suggests is a particularly appropriate strategy in a Japanese context (ibid: 173). Such approaches can invite us to see places differently and remap the spaces and places we know. Bryant et al. have discussed the value of the visual-spatial thinking involved in participatory theatre 'as a way to redevelop community spaces' (2020: 408). The importance of the sensing body is central here as a site of knowledge, learning and creation, articulating children's 'lived citizenship' (Baraldi and Cockburn, 2018), as well as expressing their hopes for the future.

As a pedagogical approach, this open and playful methodology can be an entrance point to further discussion and learning through questions like Why is this important? What do we need to do for the community to look like this, or this? and What more do we need to find out? In this way, with support from a facilitator, the work can develop students' skills of critical thinking and reflection, teamwork, problem-solving and assessing priorities – but the key element is that the starting point for the learning is the students' own questions, concerns and experience, rather than a top-down learning agenda. At the same time, the activity can feed into conversations with wider community partners about the issues raised and solutions explored. In that sense, this form of theatre can be a way to practise the 'participatory citizenship education' that Percy Smith et al. called for (2019), supporting students to engage directly in action affecting the whole community. Via this 'show and tell' combination of 'creating a performance, performing and discussing' (Roerig & Evers, 2019: 135), participatory theatre therefore has the potential to be used as a pedagogical method in school that brings students and teachers together as co-researchers and co-actors in community development.

A key benefit of the approach is that the performance work created – the world that the students have mapped – can be shown to a wider audience as a way to provoke further discussion and community learning. There is something very powerful about children's performance as an act of citizenship: Plastow points out that theatre involves '"taking space" i.e. assuming a right to place one's body meaningfully in front of an audience, and therefore asserting the value of the embodied being' (2015: 111). For marginalised groups such as children, being on stage and presenting work that demonstrates their perspective promotes their visibility as well as their capacity as actors who can effect change in that community. At the same time, the performance work they present supports children to 'map' and 'remap' their vision of the community on the stage, inviting others to see the world as they see it and respond to that vision.

Phase 1 of the Project: Exploring the Methods with Children

Our project began in October 2016 with a class of 27 Grade 6 students at an elementary school in Date City, Fukushima. The students, then in their final year, had started school a month after the 3.11 disaster, so the entire period of their schooling took place in parallel with the community's recovery. After initial meetings with the school principal and class teacher, we agreed to run a series of 90-minute theatre-based workshops, two to three times a week over a period of five weeks, culminating in a final performance on school presentation day. The workshops, which took place in the school gym, were planned and led by the research team with support from the class teacher and teaching assistant. We also met regularly with the school principal to check in about progress, and he joined in the workshops at the beginning and end of the project.

The school framed the project as an international activity, since it was led by a researcher from the UK (Lloyd Williams), and this was seen as contributing to the appeal of taking part. We were encouraged to use and teach English words and phrases as part of the work; indeed, towards the end of the project, the school principal invited us to lead some extra, one-off workshops with other classes, which focused on learning about life in the UK and exchanging questions about each other. Setting up the work this way also allowed the lead researcher to be situated as a curious outsider who didn't know about life in Fukushima and wanted to learn from the children. What this created in practice was a sense of intercultural dialogue: sharing games, experiences and ideas as part of the creative workshop process. This was also supported by the Japanese research members who could intercede to 'translate' culturally specific narratives and liaise directly between students, teachers and the lead researcher. Post-workshop team discussion proved very important in exploring and extending the themes raised during the sessions.

This cross-cultural dimension echoes Parmenter's finding that 3.11 has led to a more global outlook at schools in traditionally more 'closed' Fukushima, as the worldwide support and attention the disaster brought to the region has encouraged schools and teachers to 'rethink' their position beyond the local or regional and see themselves more as part of an international community (2012: 19). This is also reflected nationally in the Japanese Ministry of Education's Revised Elementary School Curriculum Guidelines, which outline how 'moral education' should contribute 'to the development of Japanese citizens' who, alongside learning about their own culture and traditions, will 'respect other countries, contribute to world peace and the development of the international community and the preservation of the environment, and have interest in exploring new possibilities for the future' (MEXT, 2017: 2). Published in 2017, the new programme was implemented gradually in elementary schools from 2018 and fully since 2020.

The participatory theatre methodology was very new to the staff and students, as well as the rest of the research team. While music, dancing and acting are important extracurricular activities in Japanese schools and feature regularly at open days and competitions, theatre-based pedagogy is not part of the curriculum and the creative, participatory process was less familiar to those taking part. It therefore took time to establish trust, find shared ways of working and build the children's confidence to express and explore ideas. We started every workshop with games that asked the students to use their bodies as tools of expression but also invited them to make decisions, negotiate and work collaboratively. They were encouraged to extend or develop these activities or share games of their own, and together we created a repertoire of favourite exercises that we could draw on whenever we needed a quick energiser.

After these warm-up activities, each workshop involved more extended theatre-based exercises in which the students explored what they liked about their community, what had changed during their last six years at school and what they would like to see improved in the future. The children created actions on their own or in groups and went on to develop short scenes which they showed to each other. They wrote down their ideas and responses to the work being created, sometimes on their own and sometimes in discussion with others in groups. They then went on to develop these ideas into more extended scenes, as well as creating songs, dances, poetry and letters. Throughout the process, the students were called upon to make choices and prioritise issues to help identify key 'themes' emerging from the work. In the final workshops, the research team gathered the students' work together into one continuous 15 to 20 performance piece that explored the issues the children had raised. We then spent the final week or so rehearsing, adding material as we went along to clarify or improve the piece we had created.

The final piece the children performed on presentation day was quite stylised. Structured around a series of scenes that focused on the students' key themes, the script was held together by two narrator figures who spoke directly to the audience. All the children were on stage all the time, symbolising their collective identity as a Grade 6 'community' that had grown up together through their years at school and had shared experiences to draw on in creating the piece they performed. At the same time, however, the children used the performance itself – the physical landscape of their bodies on the stage – to create a living map of the wider community beyond the school hall.

At different points in the piece, the children's bodies moved in and out of different expressions of the people and places that they saw as important in Date City: sometimes the children depicted activities that they had identified as enjoyable such as playing games with friends, relaxing or helping out at home or walking through local woodland; sometimes they embodied other people and places that they saw as important in shaping the landscape of their lives, such as friendly local people, their school or the major new road bridge that had just been opened and which meant their community was now better connected to other parts of the region. The performance also slipped between embodying the children's current view of their community and their vision of how it *could* be, presenting both their perceptions of their world in the present with their expressions of hope for the future. In this way, they mapped out a vision of the local community from a child's-eye view, high-lighting key observations they had made about their community during the workshop process and the ideas for change that they had explored.

A key theme that had emerged during the workshops was the children's awareness of the declining population and some of the effects of this. The final performance featured a song and dance piece that explored the children's call for more shops and other facilities for both younger and older people. The song began by describing the town as a 'lonely place' where the 'residents are troubled: why are there so few stores?' The song then set out the children's suggestion that investing in development would help to encourage people to the area by making it a more attractive and sociable place to live: 'Why don't you build a store with such large land? ... If there are more stores, the town will become more lively'. During the song, some of the children used their bodies to populate the stage with 'shops', demonstrating the buzz of activity that they wanted to see. Interestingly, the children also suggested that a busier town with more facilities could promote their own development by creating the opportunity to go 'shopping by ourselves'. This was shown in the dance routine when two children joined arms to gleefully 'enter' the door of one of the stores. The song ended with the children lined up on stage making a plea to 'do our best altogether!' and 'join forces!' highlighting their ambition to contribute to future planning.

This ambition was echoed in another scene in the final performance that focused on a letter some of the children had written to their local mayor about the condition of their outdoor park. This letter, which was read out loud and accompanied by actions to demonstrate the points being made, explained the children's dismay about the poor condition of the park and made a plea for it to be repaired. One child stood in the centre of the stage reading out the letter amid stark representations of the broken equipment, played by other children, twisted into contorted positions. As she called out, 'Look at this!' she invited the audience to see the park from the perspective of the children: bleak, dangerous and not a place they would want to visit.

The letter explained the children's view that having no outside place to play was potentially damaging their physical and mental health while also having negative social implications: 'Our friends are only playing video games. If we do not move our bodies, stress will pile up. ... We are completely unable to play with our friends and classmates'. Indeed, the letter stated bluntly: 'We don't know how to play'. What was being made clear here was that the park had fallen out of use for this generation of children and the space of the park – which represented play and creativity – had been remapped into a 'no-go area' that the children had little lived experience of using and no collective memory to pass on to the next generation. Interestingly, they also, like the shopping group, connected the lack of facilities to the decline in population and noted the need for facilities for all the population, not just children: 'Parks are very important to people in the community. Parks are places where everyone gathers. Many people can gather there and become friends'. As the girl reading the letter made this call for the community to recognise the value of the park as social space for local people, she role-played 'repairing' the broken equipment. As she did this, the scene changed mood as other children started gathering, smiling and playing on the swings. In this way, the scene demonstrated how the simple act of repairing the park could change the lives of young people and bring benefits to the whole community.

Another scene during the performance highlighted a recurring theme that came up during the workshops: the significance to the children of their local natural environment. One group wrote a series of *senryū* poems, highlighting their ideas about how to protect nature, including keeping rivers clean and preventing excessive logging and littering. In one powerful scene, the children demonstrated the chopping down of a tree while proclaiming, 'Stop destroying nature and excessive logging! The relaxing scent of nature, we must not destroy it'. In another, they role-played a campaign to prevent littering, some of which they attributed to decontamination workers from outside the area who they saw as lacking the same attachment to the value of the local natural environment. Here, as throughout the performance, the children's actions showed the possibility of change – whether it was the energy that comes to a town through commerce, the way that rebuilding a playground can help to rebuild a sense of community or the benefits of looking after the environment and how young people can be involved in raising awareness about this among local people.

The performance ended with a 'mirror dance', which was devised from the children's favourite warm-up activity during the workshops. It started with the children mirroring simple, slow actions in pairs, to the accompaniment of one of the students playing a piece on the piano; gradually, the pairs joined into fours and then into eights, and eventually, everyone came together to form a semi-circle, creating the same pattern of movement together. This very abstract piece of movement symbolised the power of the collective and how by working together the students could change the space of the stage – and the space of the world they were representing on that stage.

A key part of the performance came near the end when a couple of the children came off the stage and invited the special guests in the audience to respond directly in writing to the issues the performance had raised. Later, the children's parents were also invited to write and send in their responses to the piece. Feedback from the adults included acknowledgement and understanding of the issues being raised, surprise at the children's maturity and ability to engage with the issues identified and a recognition that adults have a responsibility to respond to children's suggestions and ideas. In this sense, the methods revealed the potential, as one audience member put it, for children to 'teach' adults, based on their experiences and perceptions and thus for children to have a role in contributing to community action and social change. It also highlighted how the performance had the effect of mobilising a sense of responsibility among some in the audience to respond to what they had seen and heard – in stimulating them also to 'act'.

After the performance, we asked the students to reflect on their participation in the project through written feedback and a focus group discussion with a member of the research team. Many of the students talked about the work as 'fun', but they also identified a range of learning outcomes. Several of these resonate with the kinds of critical thinking skills that Pfefferbaum identified as a benefit of involving children in participatory research on disaster risk reduction, including personal learning ('I have grown' through the project) and social learning ('our kizuna [community bond] deepened'). In addition, many of the children identified how the project had helped them learn to work together and some articulated how the learning was interdisciplinary: 'I studied many things, not just about performance. We talked about how we want the town to be'. Several students noted that they enjoyed the applause of the audience on presentation day, highlighting the value to the children of simply taking space on stage and commanding an audience.

The teachers involved in the project also talked afterwards about how the methods helped to build the students' confidence, self-expression, teamwork and independent learning. One mentioned that a student had developed his ability to use eye contact and look directly at others; another noted that some of the children went on to explore the project themes further in their other work, such as writing stories about the environment. In follow-up interviews, the teachers also discussed how the methods challenged conventional approaches in formal education settings, one saying that the children thought 'they shouldn't speak that way in front of adults' but that the work had helped them learn to speak up about their opinions. This pointed to the fact that the work helped the teachers gain experience in new pedagogical approaches around facilitating the students to express their views. Certainly, the classroom teacher became increasingly engaged in the process of delivery as the project went on, requesting session plans in advance, discussing ideas and planning with the research team and following up on activities independently.

The school principal stated that the project had been a 'fantastic encounter and good learning opportunity for the school and the region'. He went on to say that the 'spirit' the children learned of thinking on their own and expressing should be 'the basis for their education' and that he was keen to continue and expand the work by embedding it further into the curriculum. This was because the performance had made him realise that 'we need to bring the voices of children to realise their demands'. He added that 'adults need to respond to the students' requests' because 'children's development is linked to the town's development', articulating a vision for how students' learning could intersect with wider community development and highlighting a potential role for schools as hubs for that development.

Phase 2: Developing the Participatory Pedagogy with Schools

The success of the initial phase of the project and the enthusiasm of the teachers to sustain the momentum of the work led to a short series of further workshops in Date City schools a year later in 2017. From this, the team developed a new 'mini' workshop programme which could be run as a one-off intervention in other elementary schools. The mini workshops centred on two key questions: 'What do you like about living in Fukushima?' and 'What would you change to make Fukushima a better place to live?' We approached these questions through the same theatre and discussion-based methods used during the first phase: starting the workshops with warm-ups, going on to discussion and theatre-making work and ending with performance and feedback. As more schools and teachers became interested in the approach, we were invited by the Date City Board of Education to devise a teacher-training programme about our methods in order to explore further their application in educational settings in Fukushima. In this way, we began to investigate ways in which we could pursue the vision of the first school principal we worked with – where children's development is connected to the town's development.

We piloted the teacher training scheme with six teachers from Date City over a three week-period in the autumn of 2018. The programme involved first leading a workshop with elementary school children; the teachers attended and took part in this, alongside the students. Immediately after the workshop, we held a seminar for the teachers, looking at some of the theory behind the participatory theatre approach and exploring how to structure a workshop like this – as well as thinking about how the methods could be adapted. The teachers then tried running a workshop in their own school before coming back together for a final session. Here we considered the possibilities and constraints and the ways that the work could be developed further and the teachers submitted a final reflective report (Goto et al., 2020).

In 2019, we reran the now-accredited training programme, this time with 20 teachers from Date and Fukushima Cities, continuing to explore with them how this work could contribute to community resilience building. Written feedback and discussion with the teachers highlighted the potential and the challenges of the methods in schools in Fukushima. Many teachers commented, as they had during the first phase of the project, on the value of the methods in helping children to learn self-expression, creativity, teamwork and communication, as well as self-reflection 'through trying to be understood by others', as one teacher noted. These learning outcomes were identified by one training participant as life skills, 'necessary for them after going out into the world'.

Teachers reflected on the usefulness of the activities in encouraging children to express their own opinions and ideas. One noted that 'school tends to dominate children'; another commented on how the work had reinforced 'the importance of children's independence in learning'. It was again observed that some children had been inspired to continue the work on their own, with one child independently drawing pictures about 'designing the future of our town'. This was echoed by other feedback which highlighted how many of the children enjoyed the work ('They asked to do it again!') and how the students were motivated by being asked their opinion and that this encouraged participation. Certainly, the methods were identified by one teacher as an effective way for certain children to express feelings and opinions that they might find more difficult to do through other learning activities, such as writing and speaking. It's important to note that the teachers reported success in applying the methods on their own, given that one of the benefits of the first phase of the project had been its international dimension. Feedback from teachers suggests that the methods themselves could stimulate the children's self-expression and exploration of the issues, regardless of who was facilitating the activity.

There was also a range of challenges identified with the work, and these often connected with issues of facilitation. These included recognising the importance of supporting some students further to ensure that everyone's opinions are heard and included, as well as the need for teachers to gain more experience in how to facilitate this kind of work. In addition, a key question raised was about where and how to embed the work into the curriculum while also exploring the connection with the community in order to deepen the learning outcomes. This highlights the need to follow up here with more discussions with schools, teachers and local communities about what 'resilience' means in Fukushima and how children can contribute to that ongoing conversation.

One way we hope to explore this is through the new elementary school Integrated Studies curriculum in Japan, which specifically aims at engaging children in active learning about their community. The Integrated Studies programme is designed to help students make links between their formal education and 'everyday life and society' through practice-based, cross-curricular learning that is adapted to and connected with their local community context (MEXT, 2017). This new aspect of the curriculum seems an ideal space for schools to develop the kind of 'participatory citizenship education' that Percy et al. (2019) envisioned, and the participatory methods we have developed could form part of that pedagogical approach. There are exciting ways that schools could form networks with other neighbourhood groups and organisations and facilitate forms of knowledge exchange between children and their local community around present and future needs. This points to a model of 'education as democracy' as a member of the research team put it, whereby children learn through participatory pedagogical practices that help their development as active citizens, capable of contributing to resilience building, while also promoting schools as hubs for community development.

Next Steps

Teachers we have worked with have gone on to apply the methods in their own schools and the research team has developed further work with children in Japan and other countries in Asia that uses participatory theatre to elicit children's perspectives on health issues and involve them in sharing their learning with the wider community. Our team's aim was to return in 2020 and work with teachers to explore how to embed the methods into the Integrated Studies curriculum but this plan was suspended because of the COVID-19 outbreak. Instead, we went on to develop a range of 'creative health' materials for schools to use with students in response to the pandemic. This started with a plan to develop an information leaflet for COVID-19 prevention, based on local needs identified through the network of teachers involved in the participatory theatre project (Benski et al., 2020; Goto and Murakami, 2020). Our team drafted materials and revised them in collaboration with teachers and children. The developed material was then distributed to students in April 2020 and posted on school walls and websites.

In order to facilitate the distribution, we collaborated with the Municipal Office and the Board of Education. This iterative process of drafting, revising and distributing was repeated several times to produce different types of children's self-learning materials, not only about infection prevention but also to support children to carry out 'Adventures at Home' (Creative Health Team, 2020). The adventure workbook asked children to draw on a range of knowledge and skills, including mathematics, science and social studies, to conduct research about their daily living. It was noteworthy that one school led by a principal who observed the workshops in 2016 used both the workbook and the previously learned participatory theatre approach in the Integrated Studies classes. His students presented their achievements to parents on their 2020 school presentation day, using an approach that sought to 'teach' parents what they had learned and even directed questions to them (Goto et al., 2022). This example shows how the participatory theatre pedagogy, which was new to schools when applied as part of the recovery activities after the 3.11 disaster, could be adapted to the local educational context and used when facing another global health crisis.

Conclusion

Shin-Ichi Kikuchi, former president and chair of the board of trustees at Fukushima Medical University, has observed that

> [h]appiness comes from being able to have hope for the future and cherish the past. But disaster victims have lost both in a way: they block their past and are unable to see a positive future, thus creating endless anxiety. It is tragic to hear people living in evacuation shelters say, 'I am filled with so much anger that I just cannot budge an inch' (2013: 155).

Kikuchi goes on to say that it is vitally important for affected people to 'get up and move around' (ibid). This emphasis on action, on finding the capacity to act, is very important after a disaster. Having the space and opportunity to come together with others to share experiences and learn is one way of inviting people to 'get up and move around'. It can be a vital stimulus to action that supports people to reflect on the present (and the past) in a way that looks towards informing the future. It can help people re-engage with their community at an embodied and sensory level and remap the spaces in which they live according to their needs and hopes.

We suggest that the methods used on our project have provided a forum for children to explore and inform how we can do things differently in order to help build more resilient communities in Fukushima. The work has also created the opportunity for teachers to explore new teaching methods that support their students to think and act on their own. The work is ongoing, as indeed Fukushima's recovery is also ongoing, but the project has highlighted the potential for schools to take on a key role as resilience hubs in the community in Fukushima and schoolchildren as key actors in that endeavour.

References

Amri, A., Haynes, K., Bird, D. and Ronan, K. (2018) Bridging the divide between studies on disaster risk reduction education and child-centred disaster risk reduction: a critical review. *Children's Geographies*, Vol. 16 (3), pp. 239–251.

Baraldi, C. and Cockburn, T. (2018) *Theorising childhood: citizenship, rights and participation*. Basingstoke: Palgrave.

Bartos, A. (2013) Children sensing place. *Emotion, Space and Society*, Vol. 9, pp. 89–98.

Benski C., Goto A., Reich M.R., and Creative Health teams (2020) Developing health communication materials during a pandemic. *Frontiers in Communication*, Vol. 5, p. 603656.

Bryant, C., Frazier, A., Becker, B. and Rees, A. (2020) Children as community planners: embodied activities, visual-spatial thinking, and a re-imagined community. *Children's Geographies*, Vol. 18 (4), pp. 406–419.

Creative Health Team (2020). *Adventures at home, learning about life* [Online]. Available at: https://aya-goto.squarespace.com/blog/2020/7/26/childrens-workbook [Accessed: 22 March 2021].

Djohari, N., Pyndiah, G. and Arnone, A. (2018) Rethinking 'safe spaces' in children's geographies, *Children's Geographies*, Vol. 16 (4), pp. 351–355.

Geilhorn, B. (2017) Challenging reality with fiction: imagining alternative readings of Japanese society in post-Fukushima theater. In: Geilhorn, B. and Iwata-Weickgenannt, K. (eds.) *Fukushima and the arts: negotiating nuclear disaster*. Milton: Routledge, pp. 162–176.

Gibbs, L., Macdougall, C., Mutch, C. and O'Connor, P. (2017) Child citizenship in disaster risk and affected environments. In: Paton D and Johnston D.M. (eds.) *Disaster resilience: an integrated approach*. 2nd ed. Springfield: Charles C. Thomas, pp. 138–157.

Gordon, E., Elwood, S. and Mitchell, K. (2016) Critical spatial learning: participatory mapping, spatial histories, and youth civic engagement. *Children's Geographies*, Vol. 14 (5), pp. 558–572.

Goto, A., Lloyd Williams, A., Okabe, S., Koyama, Y., Koriyama, C., Murakami, M., Yui, Y. and Nollet, K.E. (2022). Empowering children as agents of change to foster resilience in community: implementing "creative health" in primary schools after the Fukushima nuclear disaster. *International Journal of Environmental Research and Public Health*, Vol. 19 (6), 3417.

Goto, A, Lloyd Williams, A., Kuroda, Y. and Satoh, K. (2020) Thinking and acting with school children in Fukushima: implementation of a participatory theatre approach and analysis of teachers' experience. *JMA Journal*, Vol. 3 (1), pp. 67–72.

Goto A. and Murakami, M. (2020) Health communication skills needed among health professionals during the COVID-19 pandemic. *Tokyo Pediatric Association Journal*, Vol 39 (2), pp. 1–5.

Itagaki, S., Harigane Maeda, M., Yasumura, S., Suzuki, Y., Mashiko, H., Nagai, M., Ohira, T. and Yabe, H. (2017) Exercise habits are important for the mental health of children in Fukushima after the Fukushima Daiichi disaster: the Fukushima Health Management Survey. *Asia Pacific Journal of Public Health*, Vol. 29 (2), pp.171S–181S.

Kikuchi, S. (2013) Special interview: tragedy to triumph. In: *Fukushima Medical University. Fukushima: lives on the line – a compendium of reports from the 2011 Great East Japan earthquake and tsunami*. Fukushima: Fukushima Medical University, pp. 152–159.

Koikari, M. (2020) *Gender, culture, and disaster in post-3.11 Japan*. London: Bloomsbury.

Lieber, M. (2017) Assessing the mental health impact of the 2011 Great Japan earthquake, tsunami, and radiation disaster on elementary and middle school children in the Fukushima Prefecture of Japan. *PloS ONE*, Vol. 12 (1), p. e170402.

Lloyd Williams, A., Bingley, A., Walker, M., Mort, M. and Howells, V., (2017) "That's where I first saw the water...": mobilizing children's voices in UK flood risk management. *Transfers: Interdisciplinary Journal of Mobility Studies*, Vol 7 (3), pp. 76–93.

Lloyd Williams, A., (2015) Exploring theatre as a pedagogy for 'developing citizens' in an English primary school. In: Prentki, T. (ed.) *Applied theatre: development*. London: Methuen, pp. 202–221.

Martz, C., Powell, R. and Shao-Chang Wee, B. (2020) Engaging children to voice their sense of place through location-based story making with photo-story maps. *Children's Geographies*, Vol 18 (2), pp. 148–161.

MEXT (Ministry of Education, Culture, Sports, Science and Technology, Japan) (2017) *Revised elementary school curriculum guidelines, English translation* [Online]. Available at: https://www.mext.go.jp/a_menu/shotou/new-cs/1417513_00001.htm [Accessed: 22 March 2021].

Mort, M., Rodríguez-Giralt, I. and Delicado, A. (2020) Introducing CUIDAR: a child-centred approach to disasters. In: Mort, M., Rodríguez-Giralt, I., and Delicado, A. (eds.) *Children and young people's participation in disaster risk reduction: agency and resilience*. Bristol: Policy Press.

Mort, M., Walker, M., Lloyd Williams, A. and Bingley, A. (2018) From victims to actors: the role of children and young people in flood recovery and resilience. *Environment and Planning C: Politics and Space*, Vol 36 (3), pp. 423–442.

Nishikawa, Y., Suzuki, C., Takahashi, Y., Sawano, T., Kinoshito, H., Vlero, E., Laurier, D., Phan, G. Nakayama, T. and Tsubokura, M. (2020) No significant association between stable iodine intake and thyroid dysfunction in children after the Fukushima nuclear disaster: an observational study. *Journal of Endocrinological Investigation*, Vol 44 (7), pp. 1491–1500.

O'Connor, P. (2015) Theatre in crisis: moments of beauty in applied theatre. In: Prentki, T. (ed.) *Applied theatre: development*. London: Methuen, pp. 185–201.

Oe, M., Maeda, M., Ohira, T., Itagaki, S., Harigane, M., Suzuki, Y., Yabe, H., Yasumura, S., Kamiya, K. and Ohto, H. (2019) Parental recognition of bullying and associated factors among children after the Fukushima nuclear disaster: a 3-year follow-up study from the Fukushima Health Management Survey. *Frontiers in Psychiatry*, Vol 10, p. 283.

OHCHR (Office of the United Nations High Commissioner for Human Rights (1989) *Convention of the rights of the child* [Online]. Available at: https://www.ohchr.org/en/professionalinterest/pages/crc.aspx [Accessed: 22 March 2021].

Ohtsuru, A., Takahashi, H. and Kamiya, K. (2019) Incidence of thyroid cancer among children and young adults in Fukushima. *JAMA Otolaryngology-Head & Neck Surgery*, Vol. 145 (8), p. 770.

Parmenter, L. (2012) Community and citizenship in post-disaster Japan: the role of schools and students. *Journal of Social Science Education*, Vol. 11 (3), pp. 6–21.

Pfefferbaum, B., Pfefferbaum, R. and Van Horn, R. (2018) Involving children in disaster risk reduction: the importance of participation. *European Journal of Psychotraumatology*, Vol 9, p. 1425577.

Percy-Smith, B., Thomas, N., Batsleer, J. and Forkby, T. (2019) Everyday pedagogies: new perspectives on youth participation, social learning and citizenship. In: Walther, A., Batsleer, J., Loncle, P. and Pohl, A. (eds.) *Young people and the struggle for participation: contested practices, power and pedagogies in public spaces*. Milton: Routledge, pp. 179–217.

Plastow, J. (2015) Embodiment, intellect, and emotion: thinking about possible impacts of Theatre for Development in three projects in Africa. In: Flynn, A. and Tinius, J. (eds.) *Anthropology, theatre, and development: the transformative potential of performance*. Basingstoke: Palgrave Macmillan, pp. 107–126.

Reich, M. and Goto, A., (2015) Towards long-term responses in Fukushima. *The Lancet*, Vol 386, pp. 498–500.

Roerig, S. and Evers, S. (2019) Theatre elicitation: developing a potentially child-friendly method with children aged 8–12. *Children's Geographies*, Vol. 17 (2), pp. 133–147.

UNDRR (United Nations Office for Disaster Risk Reduction) (2015) *Sendai framework for disaster risk reduction 2015–2030* [Online]. Geneva: UNDRR. Available at: https://www.undrr.org/publication/sendai-framework-disaster-risk-reduction-2015-2030 [Accessed: 22 March 2021].

Wisner, B., Paton, D., Alisic, E., Eastwood, O., Shreve, C. and Fordham, M. (2018) Communication with children and families about disaster: reviewing multi-disciplinary literature 2015–2017. *Current Psychiatry Reports*, Vol. 20, p. 73.

Yamamura, E. (2016) Impact of the Fukushima nuclear accident on obesity of children in Japan (2008–2014). *Economics & Human Biology*, Vol. 21, pp. 110–121.

13 Bonding after Fukushima

The Role of Trust Relationships between Non-Profit Organisation Volunteers and Disaster Victims in Building Resilience amidst a Nuclear Catastrophe

Giulia De Togni

This chapter outlines the narratives produced by a group of self-evacuated single mothers who left Fukushima Prefecture with their children after 3.11 and moved to temporary housing facilities in the Kansai region of southern Japan. These women were helped by local non-profit organisation (NPO) volunteers to rebuild their lives, acquire scientific knowledge about radiation, navigate compensation schemes and ultimately engage in the Fukushima litigation against the Tokyo Electric Power Company (TEPCO) and the state. The chapter explores the role of NPOs in empowering disaster victims through drawing attention to how activists and volunteers encouraged individuals affected by the Fukushima disaster to pursue the implementation of new rights and to make institutions accountable. New forms of grassroots politics emerged from this process, which challenged the state's authority in post-nuclear disaster Japan. The chapter ultimately suggests that NPOs can become effective mediators in helping build a dialogue between the authorities and local affected communities and improve disaster preparedness and response.

Social Capital after 3.11

In 2011, soon after 3.11, *kizuna* – meaning 'bonds' or 'connections' – was selected as the kanji (Chinese character) of the year in Japan and immediately became the slogan for the recovery campaign in the Tōhoku region. This kanji was selected to communicate a feeling that Japanese people needed to come closer together in order to overcome the worst crisis since the Second World War. In the aftermath of the Fukushima disaster, *kizuna* became a buzzword to refer to the ways in which people across Japan could support the local affected populations. The safety discourse promulgated by the state, and in particular the so-called 'Let's Eat to Support!' (Tabete ōen shiyō!) campaign (Kimura, 2016: 8), demanded that the Japanese populace consume the produce of Fukushima. This also meant trusting that the authorities were doing everything in their power to ensure that contaminated foodstuff was not made available on supermarket shelves across Japan. The *kizuna* narrative resulted in expectations and social pressure particularly in Fukushima Prefecture, where people were expected to consume the local produce (*chisan chiso*) and to silence any 'harmful rumours' (*fuhyō higai*) about radioactive contamination.

DOI: 10.4324/9781003182665-17

As a result of the *kizuna* discourse, those individuals who remained concerned – in particular pregnant women and parents of young children – were stigmatised and marginalised by their own communities, and in some cases by their own families, for spreading *fuhyō higai*. This also led to an increase in divorce rates in the affected areas – a phenomenon so widespread that a new term was coined: 'nuclear divorce' (*genpatsu rikon*). Women who divorced their husbands and self-evacuated with their children became *boshi hinansha* ('mother–child evacuees'), a sub-category of the so-called *jishu hinansha* ('voluntary evacuees'). This chapter focuses on a group of *boshi hinansha* who relocated from Fukushima Prefecture to the Kansai region. The ethnographic material analysed in this chapter is based on 15 months of ethnographic fieldwork which I carried out in Japan in 2016–2017 while collaborating with Japanese NPOs that organised recreational activities and health screenings for nuclear evacuees living in temporary housing facilities in the Kansai region.

For this research, I carried out multiple semi-structured qualitative interviews with 60 *jishu hinansha*, and for a year I accompanied 30 of them to courtrooms where they acted as plaintiffs against TEPCO and the state in the Fukushima collective civil lawsuits. To protect my collaborators' privacy, I have concealed their identities in the text, and all individual names used here are pseudonyms. The majority of my collaborators moved to Kansai from the Hamadōri region (eastern Fukushima Prefecture), in particular Minamisōma, Futaba, Namie, Iitate, Katsurao, Tomioka, and Iwaki; some of them were from the Nakadōri region (central Fukushima Prefecture), in particular Kōriyama and Fukushima City. The aim of my research was to understand how these people and their families made sense of and coped with different kinds of risks brought about by the nuclear disaster. This included not only the risk of radiation but also economic risk, job insecurity, a loss of homeland, identity crisis, state abandonment and discrimination due to widespread radiation stigma.

The research was also informed by my experience collaborating since April 2011 with Japanese NPOs involved in activities for 3.11 survivors. In particular, between 2013 and 2018, I spent six months volunteering in recreational camps (*hoyō kyanpu*) for a total of 150 children (aged 8–12) living in the most contaminated areas of Fukushima Prefecture. A wide range of *hoyō* (recuperation) projects were initiated in Japan and abroad soon after the Fukushima disaster, including *hoyō kyanpu* – which aim to ensure that the children spend a month per year in less contaminated environments. Over the past decade, these initiatives have enabled concerned parents who could not evacuate with their children to cope with the uncertainties brought by the nuclear catastrophe. As such, *hoyō kyanpu* became part of the survival practices (*sabaibaru*) adopted by parents who remained concerned that the contamination may pose a risk to the health and wellbeing of their families. Ogawa (2014: 656) points out that '*hoyō* projects are a tool to implement the right to evacuation and a step toward exploring methods of survival in the post-Fukushima era'. Notably, *hoyō* projects for Fukushima children are not sponsored by the authorities in Japan and activities such as *hoyō kyanpu* have struggled to survive because of the lack of financial aid from the state. As a result, several NPOs eventually had to cancel their *hoyō* projects owing to a lack of funds.

In the aftermath of the Fukushima disaster, some NPOs in Japan resisted forms of state authority by engaging in activities such as encouraging local residents and nuclear evacuees to self-acquire scientific knowledge on radiation and engage in the Fukushima litigation against TEPCO and the state. In this chapter, I focus on how my collaborators from Fukushima established new relationships of trust with some of these NPOs' volunteers and how this process influenced their identities as disaster 'victims' (*higaisha*) and 'neoliberal agents of change' (on neoliberalism see e.g. Greenhouse, 2010 and Ogawa, 2015). My collaborators used the word *higaisha* when talking about their experiences of the triple disaster and especially when discussing the nuclear catastrophe. Therefore, although I acknowledge the wider trend in disaster studies refusing to use the label 'victim' (e.g. Marino and Faas, 2020: 38), I use the term *higaisha* throughout the chapter, as this is a keyword important to understand how my collaborators made sense of their own lived experiences of 3.11.

Here I offer excerpts of verbal exchanges between my collaborators – both nuclear evacuees and NPO volunteers – to explore their use of emic terms (Sahlins, 2017) which refer to social capital (Aldrich, 2019). In particular, I focus on keywords such as *kizuna* (social bonds), *tsunagari* (bonding), *ibasho* (the place where one belongs) and *shinyō* (trust). Through this analysis, I show how the newly established bonds of trust between NPO volunteers and nuclear evacuees were an important factor that made my collaborators feel empowered and 'connected' (*tsunagatte iru*) and ultimately allowed them to find new strength to rebuild their lives, as well as new hope (*kibō*).

Drawing on social, legal and linguistic anthropology, this chapter builds on literature about risk, vulnerability, social capital, health and wellbeing to provide an insight into the post-Fukushima context through the point of view of some nuclear evacuees. I build on three interconnected themes or 'frictions' that resulted from the nuclear crisis: (1) mistrust towards the authorities versus trust towards NPO volunteers; (2) despair, loss, and trauma versus hope for a 'better' (imaginary) future, where nuclear victims will be legally protected; and (3) intrinsic frictions within the *kizuna* discourse due to its complexity. I highlight how NPO volunteers in Japan have often replaced the state in times of crisis, disaster and a lack of social protection. Ultimately, I conclude that the case of Fukushima in particular has shown the extent to which these groups have an empowering force which can form and transform the civil identities of disaster survivors through the creation of shared networks of trust.

The Role of NPOs in Times of Crisis

The first NPOs in Japan appeared in 1995 in the aftermath of another major disaster. On 17 January 1995, a magnitude 7.3 quake struck the Kobe area, killing more than 6,400 people, injuring over 43,000 and forcing more than

316,000 to evacuate. The then unprecedented scale of the crisis overwhelmed the authorities, who took days to organise a rescue response. A report by the city of Kobe lately documented that only about 20 per cent of those trapped in the ruins were rescued by public entities, whereas the remaining 80 per cent were rescued by untrained citizens. About 1.4 million volunteers arrived immediately in Kobe from elsewhere in Japan to aid with the rescue operations. The remarkable success of these voluntary groups, combined with the delayed response from the authorities, led to a general consensus in Japan on the need to support volunteer-based non-profit social activities (Bestor et al., 2011: 190). Following the establishment of a new law in 1998, setting up an NPO in Japan became easy, and as of 2015, there were already more than 50,000 NPOs across the country, promoting a wide range of causes and activities (see https://www.jnpoc.ne.jp/en/nonprofits-in-japan/).

Collaboration (*kyōdo*) between NPOs and the government in policy-making and policy implementation has been a popular strategy in Japan since the enactment of the 1998 NPO Law, because it has ensured effective policy implementation while saving costs for the state. This strategy has encouraged people to engage in NPOs' activities to improve the quality of life in their own communities and beyond. This process is not unique to Japan, as it aligns with global practices of neoliberalism that see citizens and not states held responsible for their own quality of life, health, and wellbeing. Within Japan's civil society (*shimin shakai*), Ogawa (2011: 192) points out that 'the state has subsequently increasingly utilized NPOs as a tool to provide services previously expected of the state'. In particular, NPOs became strong agents of neoliberalism after major crises such as the Kobe earthquake and 3.11.

After the Fukushima disaster, civil society greatly contributed to emergency relief and assistance of disaster victims in the affected regions of Tōhoku (see e.g. https://www.jstage.jst.go.jp/article/jsij/2/1/2_26/_article/-char/ja/). More than 300 Japanese NPOs and international non-governmental organisations (NGOs) mobilised their staff and emergency supplies and organised thousands of volunteers who went to north-eastern Japan to provide aid. Notably, in 2011, social media facilitated the disaster response as a gateway of information to coordinate these vast groups. In the years that followed the disaster, some NPOs continued to help parents of young children living in the most contaminated areas of Fukushima Prefecture to deal with the fear of radiation by organising medical counselling and screening sessions, as well as recreational summer/spring camps for children. Furthermore, several NPOs have supported older adult and disabled evacuees living in temporary housing facilities (*kasetsu jutaku*) by providing them with health screenings and mental health/wellbeing counselling sessions (*sōdankai*). Importantly, some NPO volunteers have also openly resisted forms of state authority by encouraging affected individuals to gain scientific knowledge on the issue of radiation and, in some cases, to engage as plaintiffs in the Fukushima litigation against both the state and TEPCO.

After the disaster, widespread mistrust towards the authorities, combined with delayed and confusing information about risk, augmented anxiety amongst the local populations. Moreover, the clash of expert-lay knowledges further contributed to fear among concerned individuals that the real dangers posed by radioactive contamination might have been downplayed in order to prioritise the economy and reconstruction efforts. In addition to this, local communities in Fukushima often lost mutual trust and social capital (Aldrich, 2019) because of discriminatory compensation schemes which differentiated between 'mandatory evacuees' (*kyōsei hinansha*) and 'voluntary evacuees' (*jishu hinansha*). Only *kyōsei hinansha* were officially recognised as victims of the Fukushima disaster and thus eligible for compensation. In this context, civil society and NPO volunteers became an important point of reference, particularly for *jishu hinansha*, providing a trustworthy network on which underrepresented and disadvantaged disaster victims could rely. The latter included single mothers with young children, older adult and disabled nuclear evacuees.

The post-Fukushima situation, in which individuals were left alone to decide where to draw the line between 'safe' and 'unsafe', is not unique to Japan. Globally, in the context of neoliberalism, risk governmentality is increasingly falling on individuals, while a sense of uncertainty and insecurity permeates citizens' discourses on identity and sense of self (Bauman, 2012 [2000]). After a nuclear disaster in particular, the cracks in the global political economy become more visible (Tsing, 2015), and it thus becomes crucial to maintain a balance between the conflicting discourses of the government and the governed (Ferguson and Gupta, 2002; Miyazaki, 2010). In such a context, NGOs and civil society groups, such as NPOs, can actively create new social platforms which may effectively mobilise resources towards sustained litigation and contribute to balancing these conflicting dialogues by empowering the powerless (Epp, 2010).

Here I focus in particular on how some NPOs have supported *jishu hinansha* to rebuild their lives through helping them find temporary housing facilities and new occupations, connect with others, navigate the compensation schemes and, finally, engage in the Fukushima litigation. An important turning point for 25,000 *jishu hinansha* was the termination of housing subsidies for self-evacuated households in March 2017. Due to this termination, low-income families such as *boshi hinansha* (single mothers and mothers with small children who evacuated from irradiated districts without their husbands) found themselves experiencing serious financial difficulties, as they had depended on housing subsidies for years after 3.11 as a 'lifeline' (*inochi zuna*). However, the termination of the housing support was in line with the Japanese government's intention to accelerate reconstruction and rebuild the economy by urging the nuclear evacuees to return to their hometowns. Those who wanted to try to retain housing support and demand compensation initiated individual and collective civil actions against TEPCO and the government.

Kizuna Narrative after Fukushima

As briefly mentioned above, the difference in evacuation rights and compensation amounts between mandatory and voluntary evacuees is substantial, ranging from a lump sum of maximum 80,000 yen per person for *jishu hinansha* (about USD 760) to a lump sum of 6 million yen per person for *kyōsei hinansha* (about USD 57,000; Matsuura, 2012: 34). As a result, most of the plaintiffs involved in the trials were *jishu hinansha*, demanding higher amounts of compensation. Since 2011, more than 12,000 Fukushima plaintiffs have filed 30 cases in different regions across Japan, from Hokkaidō to Kyūshū, against the government and TEPCO. These cases have revolved around whether the government and TEPCO, both of whom are responsible for disaster prevention measures, could have foreseen the scale of the tsunami and subsequent nuclear meltdown in Fukushima. The plaintiffs in these court cases are requesting the right to evacuate from contaminated areas where radiation levels are above 1 mSv/year; clarification of both TEPCO's and the Japanese government's liability for not preventing the nuclear disaster; adequate compensation for losses and physical and psychological damages; medical security for pregnant women, young children, older adult and disabled evacuees; regular checkups of absorbed radiation doses in the local population; and housing support and employment measures to help nuclear evacuees rebuild their lives.

After Fukushima, the Japanese government's stance has been to limit compensation only to those who can demonstrate that the damage they received has a legally sufficient causal relationship to the nuclear disaster. However, this remains problematic since it is difficult to prove a correlation between physical damage and radiation exposure, and to this day, experts disagree on the dangers posed to human health by radiation exposure. However, 'mental anguish' has officially been recognised as a form of damage due to the nuclear disaster (Dispute Reconciliation Committee, 2012: 95). Comparative epidemiological research has widely documented how disasters, whether 'natural' or man-made, negatively impact on the mental health of the affected populations (Deeg et al., 2005; Reininger et al., 2013; Fergusson et al., 2014). In particular, Neria et al. (2008) argue that the risk of PTSD (post-traumatic stress disorder) appears higher in the aftermath of man-made or techno-scientific disasters as opposed to 'natural' catastrophes. Among others, Bromet (2012) has documented how the Chernobyl disaster seriously impacted the mental health of local residents.

Similarly, after the Fukushima disaster higher levels of psychological distress were registered amongst the nuclear evacuees in comparison with other victims of 3.11, namely those affected 'only' by the earthquake and tsunami (Yasumura et al., 2012; Yabe et al., 2014; Niwa 2014; Oe et al., 2016). Some post-3.11 research studies have explored how social capital served as an important factor to build resilience after the disaster. In particular, Iwasaki et al. (2017) investigated the role of social capital in maintaining mental health amongst the residents of Futaba town, which sits only about 4 km from the crippled reactors of Fukushima Daiichi and was entirely evacuated after 3.11. Iwasaki et al. (ibid.: 405) distributed a survey amongst the 7,000 residents (2,900 households) of Futaba and received 585 responses on which they drew to suggest that high levels of social capital play 'an important role in reducing anxiety and distress'. However, since the questionnaire was, in most cases, completed by older male residents, the authors (ibid.: 417) admitted that the results were likely to be biased: 'those worse off or less connected are less likely to respond', which could have resulted in inaccurate estimates of the effects of social capital on mental health amongst disaster victims.

Several studies of disasters have argued that deeper reservoirs of social ties, what in Japanese is referred to as *kizuna* or *tsunagari*, can improve disaster preparedness, survival and reconstruction (Aida et al., 2013; Greene, 2015; Aldrich and Sawada, 2015; Gaston et al., 2016). In particular, Aldrich (2019: 178) argues that two factors drove the processes of survival and recovery after 3.11, namely networks and resilience. However, what has been neglected in this research is the complexity of social cohesion and the fact that social ties may not always be perceived as beneficial by the affected individuals. During the time I spent with my collaborators from Fukushima, I was told very often how difficult it was for them, as concerned individuals, to voice their fears and concerns with their community and, in some cases, even with their own families. Feeling marginalised, delegitimised and silenced, many of them eventually left Fukushima Prefecture. Although social cohesion may have been beneficial in those areas of Tōhoku affected 'only' by the earthquake and tsunami (Kiyota et al., 2015), the situation appears more complex when one considers the areas affected by the nuclear disaster and, in particular, those areas within or in the immediate proximity of the mandatory evacuation zone. There, local communities became internally divided due to individual differences in terms of risk perception, and due to the intrinsic complexities of the *kizuna* discourse which called for social cohesion and the silencing of concerns.

Those who found unbearable the social pressure to silence their concerns and fear of radiation eventually evacuated as *jishu hinansha*. These people often built new meaningful social bonds (*kizuna*) with NPO volunteers as well as with other nuclear evacuees. Interestingly, these 'safety' networks were thus built beyond their local communities and, in some cases, outside their own families. Indeed, it is important to highlight the complexity of the *kizuna* discourse in order to understand what happened in Fukushima, and not to take for granted that reinforcing social bonds in local communities can always implement social resilience to disaster. Social cohesion, within and beyond the affected communities, can be beneficial only to the extent to which people are given a 'voice'. Ensuring joint policy-making and a dialogue with the authorities is also key. In the section that follows, I explain how the role of NPOs in Japan, particularly after the Fukushima disaster, has been instrumental in transforming disaster victims' civil identities and helping them empower themselves. I provide three case studies to illustrate this point and a linguistic analysis of a verbal exchange between a *boshi hinansha* and an NPO volunteer.

Being Oneself, Connecting with Others and Trusting Again

I met Yukiko for the first time in October 2016. A single mother in her forties, Yukiko had moved from Minamisōma-shi to Kansai with her six-year-old daughter soon after 3.11. Thanks to a local NPO, Yukiko found temporary housing in a flat in southern Osaka. She then joined the NPO as a volunteer, and when we met, she was working in a café owned by the organisation. During our chats at the café and in her temporary housing facility, Yukiko often mentioned how she had felt discrimination from her own community and ex-husband in Fukushima. Conversely, she felt supported by this NPO and had made meaningful bonds (*tsunagari*) with other volunteers (who often were evacuees like her). In Yukiko's words,

> This place for me is what in Japanese we call *ibasho*, you know? It is a place where I can be myself and I feel accepted by others. A place where I can find peace of mind (*anshin*).

Keiko, another *boshi hinansha* in her forties who had moved from Iitate-mura to Kansai with her four children, used similar words to describe how a local NPO had helped her rebuild her life. When we met in November 2016, Keiko was also working at a café owned by another NPO in southern Kyoto. During our conversations at her temporary housing facility and at the café, Keiko explained to me that this NPO had helped her find her flat and had offered her a job as waitress at the café. Like Yukiko, also Keiko felt that working at this café helped her connect with others:

> I am glad to be part of this NPO and to be working at this café. This is a rewarding occupation which also allows me to co-organise and take part in several volunteering activities for Fukushima children and the elderly evacuees. I feel connected (*tsunagatte iru*) to the rest of the community of nuclear evacuees.

Both Yukiko and Keiko had found an *ibasho* in these cafés owned by local NPOs. In November 2016, Keiko invited me to meet the woman who started the NPO she had joined. Chiyo, a mother in her fifties, joined us for lunch at the café in Kyoto and explained to me how she had started volunteering well before 3.11 and eventually decided to found an NPO for nuclear evacuees. After majoring in psychology, Chiyo had volunteered mainly for older adults living alone prior to 3.11 by, for instance, bringing lunchboxes (*obentō*) to their houses and providing free mental health counselling. After 3.11, she became particularly concerned about the mental health of children who had lost their parents to the tsunami. Chiyo started regularly visiting an elementary school in Ishinomaki, Miyagi Prefecture, where she provided free mental health counselling to young children who had lost one or both of their parents. She then became interested in supporting disadvantaged single mothers with young children living in temporary housing facilities as a result of the nuclear disaster:

> The situation for mothers like Keiko who moved to Kansai from Fukushima is particularly complicated. First there is the problem of communication (*kotoba*), because the Tōhoku and Kansai dialects are

different, and in Kyoto those who have a Tōhoku accent tend to be marginalised. This affects children in schools, for example, who end up being bullied due to the stigma of radiation as well. Then there is the problem of cultural difference (*bunka no chigai*) where, again, Tōhoku and Kansai greatly differ. And, last but not least, are the financial problems that single mothers in particular face. I opened this café to provide an *ibasho* for women like Keiko, a place where they could meet and be themselves, not being afraid of voicing their fears and concerns.

Keiko felt inspired by Chiyo to co-found with other *boshi hinansha* their own NPO. When we met in 2016–2017, Keiko's NPO was organising activities to support concerned parents living in the most contaminated areas of Fukushima. Such activities included the opportunity to spend a couple of weeks per year in Kansai and receive health and mental health counselling. Another of my collaborators, Kumiko, was amongst the members of Keiko's NPO. I met Kumiko first in November 2016 and, during our frequent meetings in the months that followed, she explained how her friendship with Keiko and their NPO's activities had brought new hope (*kibō*) in her life. Kumiko was also a *boshi hinansha* in her forties who had moved from Fukushima City to Kansai with her ten-year-old daughter after 3.11. She was not divorced, but her husband could not leave Fukushima for occupational reasons; thus, Kumiko was living alone with her daughter in a temporary housing facility in Kyoto.

Keiko and Kumiko were very committed to their NPO's activities and invited me to join them often for counselling sessions and also when they were discussing with other local NPOs about potential collaborations. During one of the meetings I joined in May 2017, Keiko and Kumiko met with nine volunteers of another NPO based in north-eastern Kyoto. These were all women in their forties to sixties and they promptly agreed to help Keiko and Kumiko fund their projects mainly through co-organising events such as flea markets (*furī māketto*) to collect money. This was a common strategy adopted by other NPOs with whom I had collaborated as a volunteer, which were also organising *hoyō* projects. Given the lack of financial support from the state, these NPOs were relying on such kinds of activities and private donations to carry on with their projects.

Saki, another *boshi hinansha* in her forties who had left Iwaki-shi and moved to Kansai after 3.11 with her two daughters, was also relying on such activities to support her organisation – which, when we met in 2017, had yet to be officially given the status of NPO. Saki – like Yukiko, Keiko and Kumiko – had found temporary housing in an apartment in Kansai with the help of a local NPO. As a single mother, Saki was struggling financially, juggling two part-time jobs. However, during our conversations, Saki often referred to how she had found new purpose and hope in the activities of her organisation, which was specifically meant to support *boshi hinansha* like her. Inspired by Keiko's and Kumiko's activities, Saki and the other *boshi hinansha* involved in this organisation had applied to have their group officially recognised as an NPO, but they were struggling to make this happen. In Saki's words,

> I phoned the Fukushima Prefectural office to ask why our application was rejected, and a secretary just told me: 'Your activities do not qualify because other NPOs are better than yours'. Those words hurt, you know. I asked how we could improve our application and we re-submitted it, but we have not heard back from them yet. At the moment, we are self-funding our activities. I sincerely believe that what we do is meaningful and I will continue to endure. Our association is based on human connections (*tsunagari*) and I feel supported by this network. This is the place where I am supposed to be. It's a relief to have this safe space and social connections (*ibasho to tsunagari ga atte anshin da*).

Eventually, Keiko, Kumiko and Saki all became involved as plaintiffs in the collective civil lawsuits against TEPCO and the state, and they invited me to join them in the courtroom. I spent one year attending the proceedings in a court in Kansai, where these women and other nuclear evacuees requested compensation and the right to evacuate. What follows is a verbal exchange between another *boshi hinansha*, Saeko, and a volunteer of a local NPO, Akari. I collected this conversation in court, and what emerges is how women like Keiko, Kumiko, Saki and Saeko went through a transformation of their civil identities after the Fukushima disaster and felt supported and motivated by NPO volunteers like Chiyo and Akari. Notably, these actors were all women. Women who became politically active after the disaster often identified themselves as housewives and mothers to justify their activism within the roles prescribed for them by the modern nation state (Koyama, 2014: 85).

Hope for a Better Future

Saeko is a *boshi hinansha* in her forties. She left Minamisōma with her two children soon after the three explosions at the Fukushima Daiichi nuclear power plant. An NPO had helped her to move to a temporary housing facility in Kansai. The founder of this NPO was Akari, another mother, also in her forties. Akari helped Saeko to find a new school for her daughters and a part-time job in Kansai. She also actively encouraged Saeko to acquire scientific knowledge on the issue of radiation by inviting her to join a Citizen Radiation Monitoring Station (CRMS) which had been funded by a partner NPO. When we met in 2017, Saeko was volunteering at this CRMS once a week and attending regular meetings together with other volunteers and radiation experts to measure samples of food and dust from different localities of Fukushima Prefecture. Akari also encouraged Saeko to take part in the Fukushima litigation against TEPCO and the Japanese government. The following excerpt is a conversation between Saeko and Akari which I collected at the courthouse in 2017, during a break between the hearings:

SAEKO: If it weren't for Akari san, I would not have been able to do all these things.

AKARI: You have done everything by yourself. We have just encouraged you to persevere.

SAEKO: I felt alone in my struggles after the disaster, until I found you Akari-san. In Fukushima, I felt I could trust nobody – neither the authorities' discourse on safety nor the experts who, we feared, may have been paid by the nuclear industry to reassure the local population. I had to decide what to do for my daughters, what was the best choice for them. When my ex-husband insisted that I and the girls moved back there, I refused. But I would have not found the strength to refuse this without your support – you know that Akari-san.

AKARI: You must never feel alone. There are many other evacuees and volunteers here. We have created a network that connects (*tsunagaru*) people together.

SAEKO: That's true. We are all connected (*tsunagatteiru*). I think that I found in our network the kind of hope (*kibō*) that I had lost after the disaster. I formed an important bond (*kizuna*) with you and the other volunteers here in Kansai.

AKARI: Social bonds have become especially important after 3.11, perhaps more important than ever before. There are several problems that affect nuclear evacuees, particularly in western Japan: social discrimination, radiation stigma, language and cultural differences. Integration is difficult for someone from Tōhoku, especially if they are from rural Fukushima. But we want to change that. It is also crucial that we ensure that the victims of 3.11, as well as other future disaster victims, are protected by the law. The main purpose of the Fukushima litigation is for the nuclear evacuees to gain legal recognition of their status as 'nuclear refugees' (*genpatsu nanmin*). We need the government to create new policies for disaster victims. The real problem here, what has come into light after 3.11, is that there is currently no law in Japan specifically for the victims of natural and man-made disasters. We urgently need to change that.

SAEKO: That's true. That's why we are investing our energies in these trials. I have learnt so much about my rights from Akari-san and other NPO members here in Kansai, now I want to fight for my daughters and future generations. I want them to be legally protected in case another disaster happens. Fukushima can become a turning point for Japan if we hang together in our battles. I and the other evacuees attending the trials have hope - hope for a better future.

Hage (2003: 17) argues that 'capitalist societies are characterised by a deep inequality in their distribution of hope, and when such inequality reaches an extreme, certain groups are not offered any hope at all'. The increasingly unequal distribution of hope following neoliberal reform have resulted in what Hage (ibid.: 20) terms a 'shrinking society'. In particular, Miyazaki (2010: 239) draws attention to 'the loss of hope' in Japanese society. Recently, the previously prevalent view of Japanese society as comprising 'a hundred million middle-class people' (*ichioku so churyu*) has been replaced by the view of a 'stratified society' (*kakusa shakai*), divided into 'winners' (*kachigumi*) and 'losers' (*makegumi*; Miura 2005). In this discursive shift from an emphasis on equality to an emphasis on competition, hope has emerged as a subject of intense public debate, especially after 3.11. Underlying this interest in hope and the lack of hope, is Japanese intellectuals' shared concern with the pervasive celebration by the government, as well as by the media, of the neoliberal ideal of 'strong individuals' (*tsuyoi kojin*), ready to take risks (*risuku*) while taking responsibility (*jiko sekinin*) for their own risk-taking actions.

In analysing the nexus of neoliberal economic reforms and the loss of hope, Hage, Genda, and Yamada propose different ways in which hope could be restored in society (Miyazaki ibid.: 240). Hage (2003) proposes regime change. In contrast, Genda (2005, 2006) and Yamada (2004) propose more limited and pragmatic measures, specifically targeting those whom they see as losing hope. Yet Hage, Genda, and Yamada all agree that hope no longer exists as it used to. More specifically, they all share a view that hope was more evenly distributed in societies in the past. In other words, underlying these public intellectuals' diagnosis of neoliberalism through the analytics of hope is a particular temporal frame, that is the 'temporality of no longer' (Miyazaki ibid.: 240). In his best-selling book *Exodus in a Country of Hope*, the Japanese novelist Murakami Ryū (2002: 314) writes: 'In this country, there is everything. Indeed, there are all kinds of things. The only thing that [Japan] does not have is hope'.

Nevertheless, in Saeko's narrative – as well as in the narratives of the other *boshi hinansha* reported in this chapter – the nuclear evacuees found new hope in newly established relationships of trust with NPO volunteers. In particular, *kizuna* is a term that my interlocutors often used to define their 'being intertwined' and which embodies strong nuances of embracing solidarity in one's community. In Japanese, 'radicals' are graphemes which express the general nature of the kanji (Japanese characters) and provide clues to the kanji's origin, group, meaning(s) or pronunciation(s). The radical of the kanji 絆 (*kizuna*) is 糸 (*ito*) which hints at the meaning of the kanji by bringing in a visual image – that of a thread. The term ultimately means that one is never completely alone, as everyone is connected to at least someone else. The narratives collected here all suggest that the need for others was reaffirmed after 3.11, as NPO actors and nuclear evacuees like Akari and Saeko connected and resisted forms of state authority and dominant discourses about safety and through this process found new hope.

The experience of the post 3.11 context offers a means to understand the complexity of the discourse processes by which actors in NPOs may form and transform the civil identities of disaster survivors through the creation of shared networks of trust (*shinyō*). In particular, the narratives presented here highlight how *kizuna* between volunteers and nuclear evacuees have played a significant role in bringing the grassroots voices of disadvantaged groups such as *boshi hinansha* to the access of law-making processes in post-nuclear disaster Japan. I have drawn attention to the changing forms of empowerment and disempowerment of disaster victims who have coped with the uncertainty brought by contemporary politics of neoliberalism after 3.11. The case of Fukushima is exemplary in showing the role played by NPOs in helping nuclear evacuees to build social resilience and develop new agencies to tackle the uncertainty brought about by the nuclear disaster, regaining hope thanks to *kizuna*.

Vulnerability, Resilience and Empowerment

The case studies discussed in this chapter illustrate how NPOs can support nuclear disaster victims by helping them rebuild their lives and encouraging them to empower themselves. After 3.11, civil society groups have played a considerable role in mobilising the voices of the underrepresented nuclear evacuees through actively serving as facilitators of dialogue between the evacuees and government officials from the Reconstruction Agency and through supporting the Fukushima plaintiffs in court against TEPCO and the state. Over the past decade, a number of NPOs in Japan have continued to organise gatherings for nuclear evacuees and invited officials in an attempt to initiate a constructive dialogue between the two parties. In particular, within a year after 3.11, in July 2012, a citizen network called 'Citizens' Forum for the Nuclear Disaster Victims' Support Act' was created and, as of June 2014, 64 civil society groups across Japan were involved. These included the Fukushima Network for Saving Children from Radiation, the Citizens' Radioactivity Measuring Stations (CRMSs), the Save Fukushima Children Lawyers' Network, Friends of the Earth Japan, Peace Boat and Human Rights Now. All these organisations have worked to bring the voices of those affected by the disaster to the wider public to ensure that affected individuals can participate in and actively influence the policy-making processes of post-nuclear disaster Japan. Specifically, this is required in Article 5(3) of the Nuclear Disaster Victims' Support Act which states (cited in Ogawa, 2014: 655), 'In formulating the basic policy, the government shall take measures necessary to reflect in their contents the opinions of the residents of the areas affected by the TEPCO Nuclear Accident and those who have been evacuated from the affected areas'.

Prior to the Fukushima disaster, local communities had largely ignored the risk of hosting nuclear power stations in disaster-prone areas; both TEPCO and the government had promulgated the 'myth of nuclear safety' (*anzen shinwa*) and insisted on 'zero risk'. As Aldrich (2013: 263) points out, 'Fukushima has created a situation in which citizens – after being told for decades that accidents were impossible and hearing the "safety myth" proclaimed by politicians, engineers, and public relations experts – largely distrust their leaders'. A gap in trust has formed between civil society and the state, also caused by a number of scandals and incidents involving the nuclear industry prior to 3.11. Moreover, no adequate disaster management plan was prepared in case of a nuclear accident to inform the hosting communities about risk, to prepare them for a possible disaster and adequately respond in the aftermath of this. As documented in the

Official Report of the Fukushima Nuclear Accident Independent Investigation Commission (NAIIC 2012), while the authorities had not pushed against stricter tsunami countermeasures as recommended by experts' reports, TEPCO itself had no adequate emergency plans in place and its managers and employees were not sufficiently trained to handle a blackout.

Coastal Tōhoku was an economically deprived area already prior to 3.11, with most jobs in fishing, farming and agriculture. The region was already coping with financial vulnerability and a rapidly ageing population (as elsewhere in rural Japan), since thousands of residents had left their villages to seek university education and jobs with higher salaries in larger cities, particularly Tokyo. Coastal communities in Tōhoku, which were the ones most affected by 3.11, lacked strong economic infrastructures and were particularly suffering from depopulation. These challenges are amplified in the area of coastal Fukushima that hosts the crippled reactors, as the only industries remaining after 3.11 are in the field of nuclear power, in particular dealing with the decommissioning process, reprocessing and removal of radioactive waste. The few residents who moved back there after 3.11, mainly older adults, commute in trains that run past hundreds of thousands of radioactive waste bags, piled one atop the other. Repopulating and revitalising these areas will be a challenge for decades to come.

Japan, as well as other nations that face major disasters, will need to address these issues and work towards building beneficial social ties amongst local communities – creating trust and solidarity amongst residents – and enhancing more effective governance that includes local communities in the policy-making process. These actions could make a considerable difference when facing future disasters, especially if deprived and marginalised populations are adequately informed about risk, protected after a disaster, included in policymaking, and empowered to rebuild their lives. What the case studies analysed in this chapter have shown is the potential of NPOs to form and transform the civil identities of disaster victims through the positive creation of shared networks of trust. From these narratives, Japanese citizens appear to have little sense of trust left in their government, especially amongst those I interviewed while in the field, who expressed a widespread feeling of being ignored by those in power who are supposed to represent them. It will be necessary to address this issue and rebuild citizens' trust in the authorities to ensure better disaster governance.

Fukushima has opened up policy windows with the potential for change in several fields. In this chapter, drawing on the example of some of my collaborators who engaged in activism and the Fukushima litigation after 3.11, I have addressed specifically the potential of NPOs in empowering the 'powerless' through supporting *jishu hinansha* in their struggles to be heard, bringing them to the policy-making processes and helping them succeed in claiming their right to evacuation (*hinan suru kenri*) and, ultimately, supporting them to rebuild their lives. Kimura (2016) rightly interprets the proliferation of NPOs in Japan not as the strengthening of democracy but as a tangible sign of neoliberalism.

Transformed Civil Identities

In the post-3.11 context, NGOs and civil society groups have created new social platforms that have mobilised resources to conduct sustained litigation and balance conflicting narratives between the government and the governed, through empowering nuclear evacuees and building social resilience amongst those who are powerless. To prepare for future crises, what is most needed is better governance, more inclusive policy-making and an open and more constructive dialogue between the affected populations and the authorities. NPOs seem to be ideal facilitators in promoting such dialogue, as newly created 'social bonds' (*kizuna*) between NPO volunteers and Fukushima evacuees have played a significant role in bringing the grass-roots voices of some of the poorest and most marginalised groups to the access of law-making processes in post-3.11 Japan.

The relationships of trust between NPO volunteers and nuclear evacuees have helped some of the affected individuals find new hope after an unprecedented catastrophe. In this chapter, I documented how self-evacuated single mothers (*boshi hinansha*) in particular were supported by NPOs to find temporary housing facilities, new occupations, volunteering opportunities and a sense of connection with one another. NPO volunteers have actively encouraged these women to educate themselves on the issue of radiation and have provided them with medical counselling and emotional support. And, in some cases, they have helped these single mothers to navigate the Fukushima litigation. The excerpts analysed in this chapter highlight my collaborators' use of emic terms such as *kizuna*, *tsunagari* and *ibasho* which all refer to how newly created bonds of trust between volunteers and nuclear evacuees were perceived as vital for my collaborators after 3.11. The relationships of mutual trust they created with these NPO volunteers led, in some cases, to the creation of new NPOs by the *boshi hinansha* themselves.

What clearly emerges from these narratives is the important position of NPOs in Japan, which have played a fundamental role in times of disaster, uncertainty, crisis, vulnerability and a lack of social protection since they were first created in the aftermath of the Kobe earthquake in 1995. The case of the Fukushima disaster shows the extent to which these groups can inform and transform the civil identities of disaster victims and empower the most vulnerable and disadvantaged amongst them through the creation of shared networks of trust and affective bonds.

References

Aida, J., Kawachi, I., Subramanian, S.V. and Kondo, K. (2013) Disaster, social capital, and health. In: Kawachi, I., Soshi, T. and Subramanian, S. (eds.) *Global Perspectives on Social Capital and Health*. Berlin: Springer, pp. 87–122.

Aldrich, D.P. (2013) Rethinking civil society-state relations in Japan after the Fukushima accident. *Polity*, Vol. 45 (2), pp. 249–264.

Aldrich, D.P. and Sawada, Y. (2015) The physical and social determinants of mortality in the 3.11 tsunami. *Social Science and Medicine*, Vol. 124, pp. 66–75.

Aldrich, D.P. (2019) *Black Wave: How Networks and Governance Shaped Japan's 3/11 Disasters*. Chicago: University of Chicago Press.

Bauman, Z. (2012) [2000] *Liquid Modernity*. Cambridge: Polity Press.

Bestor, V., Bestor, T.C. and Yamagata, A. (2011) *Routledge Handbook of Japanese Culture and Society*. London: Routledge.

Bromet, E.J. (2012) Mental health consequences of the Chernobyl disaster. *Journal of Radiological Protection*, Vol. 32 (1), p. N71.

Deeg, D.J.H., Huizink, A.C., Comijs, H.C. and Smid, T. (2005) Disaster and associated changes in physical and mental health in older residents. *European Journal of Public Health*, Vol. 15, pp. 170–174. doi: 10.1093/eurpub/cki126.

Dispute Reconciliation Committee for Nuclear Damage Compensation. (2012) *Japan's compensation system for nuclear damage as related to the TEPCO Fukushima Daiichi nuclear accident*. OECD and NEA (Nuclear Energy Agency). Last accessed at https://www.oecd-nea.org/law/fukushima/7089-fukushima-compensation-sysem-pp.pdf [Accessed: February 22, 2022]. 247pp.

Epp, C. (2010) *Making Rights Real: Activists, Bureaucrats, and the Creation of the Legalistic State*. Chicago: University of Chicago Press.

Ferguson, J. and Gupta, A. (2002) Spatializing states: toward an ethnography of neoliberal governmentality. *American Ethnologist*, Vol. 29 (4), pp. 981–1002.

Fergusson, D.M., Horwood, L.J., Boden, J.M. and Mulder, R.T. (2014) Impact of a major disaster on the mental health of a well-studied cohort. *JAMA Psychiatry*, Vol. 71 (9), pp. 1025–1031.

Gaston, S., Nugent, N., Peters, E.S., Ferguston, T.F., Trapido, E.J., Robinson, W.T. and Rung, A.L. (2016) Exploring heterogeneity and correlates of depressive symptoms in the Women and Their Children's Health (WaTCH) Study. *Journal of Affective Disorders*, Vol. 205, pp. 190–199.

Genda, Y. (2005) *A Nagging Sense of Job Insecurity: The New Reality Facing Japanese Youth*. Trans. Jean Connell Hoff. Tokyo: International House of Japan.

Genda, Y. (2006) *Kibōgaku* [Hope Studies]. Tokyo: Chuokoron Shinsha.

Greene, G. (2015) Resilience and vulnerability to the psychological harm from flooding: the role of social cohesion. *American Journal of Public Health*, Vol. 105 (9), pp. 1792–1795.

Greenhouse, C.J. ed. (2010) *Ethnographies of Neoliberalism*. Philadelphia: University of Pennsylvania Press.

Hage, G. (2003) *Against Paranoid Nationalism: Searching for Hope in a Shrinking Society*. Annandale: Pluto Press Australia.

Iwasaki, K., Sawada, Y. and Aldrich, D.P. (2017) Social capital as a shield against anxiety among displaced residents from Fukushima. *Natural Hazards*, Vol. 89, pp. 405–421. doi 10.1007/s11069-017-2971-7.

Kimura, H.A. (2016) *Radiation Brain Moms and Citizen Scientists: The Gender Politics of Food Contamination After Fukushima*. Durham: Duke University Press.

Kiyota, E., Tanaka, Y., Arnold, M. and Aldrich, D.P. (2015) Elders leading the way to resilience. *World Bank conference paper series* [Online]. Available at: https://www.gfdrr.org/en/publication/elders-leading-way-resilience. [Accessed: 16 July 2021].

Koyama, S. (2014) Domestic roles and the incorporation of women into the nation-state: the emergence and development of the 'good wife, wise mother' ideology. Translated by Vera Mackie. In: Germer, A., Mackie, V. and Wohr, U. (2014) *Gender, Nation and State in Modern Japan*. London: Routledge, pp. 85–100.

Marino, E.K. and Faas, A.J. (2020) Is vulnerability an outdated concept? After subjects and spaces. *Annals of Anthropological Practice*, Vol. 44 (1), pp. 33–46.

Matsuura, S. (2012) *Japan's compensation system for nuclear damage as related to the TEPCO Fukushima Daiichi nuclear accident* [Online]. OECD and NEA (Nuclear Energy Agency). Available at: https://www.oecd-nea.org/law/fukushima/7089-fukushima-compensation-system-pp.pdf [Accessed: 27 January 2021].

Miura, A. (2005) *Karyu shakai: aratana kaiso shudan no shutsugen* [The low incentive society: the emergence of a new class]. Tokyo: Kobunsha.

Miyazaki, H. (2010) The temporality of no hope. In: C.J. Greenhouse (ed.) *Ethnographies of Neoliberalism*. Philadelphia: University of Pennsylvania Press, pp. 238–250.

Murakami, R. (2002) [2000] *Kibō no kuni no ekuzodasu* [Exodus in a country of hope]. Tokyo: Bungeishunju.

NAIIC (National Diet of Japan Fukushima Nuclear Accident Independent Investigation Commission). (2012) *The official report of the Fukushima nuclear accident independent investigation commission, executive summary* [Online]. Available at: https://www.nirs.org/wp-content/uploads/fukushima/naiic_report.pdf [Accessed: 16 July 2021].

Neria, Y., Nandi, A. and Galea, S. (2008) Post-traumatic stress disorder following disasters: a systematic review. *Psychological Medicine* 38, pp. 467–480.

Niwa, S. (2014) A new structure for mental health and welfare in the Soso area to promote the recovery of people in Fukushima from the 3.11 earthquake and nuclear power plant accident. *Seishin Shinkeigaku Zasshi* 116, pp. 621–625.

Oe, M., Fujii, S., Maeda, M., Nagai, M., Harigane, M., Miura, I., Yabe, H., Ohira, T., Takahashi, H., Suzuki, Y., Yasumura, S. and Abe, M. (2016) Three-year trend survey of psychological distress, post-traumatic stress, and problem drinking among residents in the evacuation zone after the Fukushima Daiichi Nuclear Power Plant accident [The Fukushima Health Management Survey]. *Psychiatry and Clinical Neurosciences*. 70, pp. 245–252. doi: 10.1111/pcn.12387.

Ogawa, A. (2011) The new prominence of the civil sector in Japan. In Bestor, V. et al. (eds.) *Routledge handbook of Japanese culture and society*. London: Routledge, pp. 186–197.

Ogawa, A. (2014) The right to evacuation: the self-determined future of post-Fukushima Japan. *Inter-Asia Cultural Studies*, 15(4), pp. 648–658. doi: 10.1080/14649373.2014.977516.

Ogawa, A. (2015) *Lifelong Learning in Neoliberal Japan: Risk, Community, and Knowledge*. Albany: SUNY Press.

Reininger, B.M., Rahbar, M.H., Lee, M., Chen, Z., Alam, S.R., Pope, J. and Adams, B. (2013) Social capital and disaster preparedness among low income Mexican Americans in a disaster-prone area. *Social Science and Medicine* 83, pp. 50–60. doi: 10.1016/j.socscimed.2013.01.037.

Sahlins, M. (2017) In anthropology, it's emic all the way down. *HAU Journal of Ethnographic Theory* 7(2), pp. 157–163. doi: 10.14318/hau7.2.020.

Tsing, L.A. (2015) *The Mushroom at the End of the World: On the Possibility of Life in Capitalist Ruins*. Princeton: Princeton University Press.

Yabe, H., Suzuki, Y., Mashiko, H., Nakayama, Y., Hisata, M., Niwa, S., Yasumura, S., Yamashita, S., Kamiya, K. and Abe, M. (2014) Psychological distress after the Great East Japan Earthquake and Fukushima Daiichi Nuclear Power Plant accident: results of a mental health and lifestyle survey through the Fukushima Health Management Survey in FY2011 and FY2012. *Fukushima Journal of Medical Science* 60, pp. 57–67. doi: 10.5387/fms.2014-1.

Yamada, M. (2004) *Kibo kakusa shakai* [The stratified hope society]. Tokyo: Chikuma shobo.

Yasumura, S., Hosoya, M., Yamashita, S., Kamiya, K., Abe, M., Akashi, M., Kodama, K. and Ozasa, K. (2012) Study protocol for the Fukushima health management survey. *Journal of Epidemiology* 22(5), pp. 375–383. doi: 10.2188/jea.je20120105.

14 Fukushima's Reconstruction after the Nuclear Accident and the Fukushima Daiichi Nuclear Power Plant (1F) Decommissioning Policy

Shunji Matsuoka

What Lessons Have We Learned from Fukushima?

Eleven years have passed since the Great East Japan Earthquake and Fukushima nuclear accident on March 11, 2011. Following the Fukushima nuclear accident, which had such a great impact not only on Japanese society, but on the rest of the world, what new knowledge or lessons have we drawn from the recovery process regarding nuclear disasters and reconstruction?

From the Three Mile Island (TMI) accident in 1979, the world learned the importance of 'defense in depth' in preparing for severe accidents at nuclear power plants. From the Chernobyl accident in 1986, we learned the importance of nuclear safety culture (Murakami, 2005). However, the Fukushima nuclear accident in 2011 occurred because we in Japan had not learned these lessons of defense in depth from TMI and of safety culture from Chernobyl.

If the lesson to be learned from Fukushima is the difficulty in learning such lessons from history, then showing how we can learn these historical lessons and pass them on to future generations is an incredibly knowledge to acquire. Moreover, if we can resuscitate local communities from a Level 7 nuclear disaster (which had happened for the first time in an affluent, democratic society) and form sustainable communities in Fukushima, this could be a lesson in post-accident management more widely and a significant academic contribution to human society (Matsuoka, 2017).

Whether Fukushima will succeed in its efforts to revitalize the region from the Great East Japan Earthquake and nuclear disaster to form sustainable communities which do not rely solely on the nuclear decommissioning industry depends on whether new knowledge can be gained through the collaboration of various kinds of expert and local knowledge (Matsuoka et al., 2013). It also depends on whether social innovation can be created based on that new knowledge (Matsuoka, 2018). Say we name this new knowledge, gained from the reconstruction process in Fukushima, 'Fukushima reconstruction knowledge'. Clarifying what Fukushima reconstruction knowledge is will be an important point in approaching the core of the lessons to be learned from Fukushima (Matsuoka, 2020b).

This chapter focuses on the relationship between the decommissioning at Fukushima Daiichi nuclear power plant (hereinafter referred to as 1F) and the

DOI: 10.4324/9781003182665-18

reconstruction of Fukushima and considers what lessons there are to be learned. The reconstruction of Fukushima is the first undertaking in human history to tackle the dual challenges of (1) recovering from a natural disaster, namely earthquake and tsunami, and (2) recovering from a nuclear disaster; it also faces the extremely heavy task of balancing reconstruction and decommissioning.

The structure of this chapter is as follows: First, we consider the significance and background of the emergence of issues unique to Fukushima, in the formation process of 1F decommissioning policy as well as reconstruction that is compatible with decommissioning. Next, we consider the difficulties in forming social consensus regarding the management and disposal of treated water (now, the biggest issue facing the current 1F decommissioning policy), from the perspective of collaboration between expert and local knowledge. Finally, we will consider how balance can be achieved between reconstruction and decommissioning in Fukushima, from a social innovation approach, which involves making the 1F decommissioning process a local resource. This can occur through collaboration between expert and local knowledge.

1F Decommissioning Policy and the Coexistence of Reconstruction and Decommissioning

Over ten years have passed since the 1F accident in March 2011, and the 1F decommissioning work is in full swing. In order to reflect on potential lesson learning, I would first like to summarize the formation and development of the decommissioning policy for 1F over the past ten years.

Formation and Development of 1F Decommissioning Policy

Table 14.1 shows the process of the formation of the 1F decommissioning policy in three phases. The first phase is the formation of basic policy (2011–2012), which covers the formulation of the first Mid-and-Long-Term Roadmap (IMCCWD, 2019), which is the basic plan for the 1F decommissioning policy, up until its first revision. The second phase comprises the formation of the decommissioning system (2013–2014) up to the second revision of the Mid-and-Long-Term Roadmap and the establishment of the Inter-Ministerial Council, the Team for Countermeasures for Decommissioning and Contaminated Water Treatment, and the Fukushima Council for 1F decommissioning. The third phase is the operation period for the decommissioning system (2014-present), of which the third and fourth revisions of the Mid-and-Long-Term Roadmap and the establishment of the Nuclear Damage Compensation and Decommissioning Facilitation Corporation (NDF), as well as a reserve fund system, are important parts.

Mid-and-Long-Term Roadmap and the Coexistence of Reconstruction and Decommissioning

The Mid-and-Long-Term Roadmap towards the 1F Decommissioning (hereinafter referred to as the 'Mid-and-Long-Term Roadmap'), stipulates the basics of the national government and TEPCO's decommissioning policy for 1F.

Table 14.1 Formation Process of 1F Decommissioning Policy and Decommissioning Systems

Stage of Development for Policy/System	Date	Details
Formation of Basic Policy (2011–2012)	12/21/2011	1st draft of the Mid-and-Long-Term Roadmap
	7/30/2012	1st revision of the Mid-and-Long-Term Roadmap
	6/27/2013	2nd revision of the Mid-and-Long-Term Roadmap
	8/19/2013	Leakage of approximately 300 m^3 of contaminated water from storage tank
	9/3/2013	Basic Policy for the Contaminated Water Issue is decided Establishment of the Inter-Ministerial Council for Contaminated Water and Decommissioning Issues (IMCCWD)
Formation of Decommissioning System (2013–2014)	9/10/2013	Establishment of the Team for Countermeasures for Decommissioning and Contaminated Water Treatment under the Nuclear Emergency Response Headquarters
	12/20/2013	Additional Measures for Decommissioning and Contaminated Water Issues decided Council for Decommissioning is integrated into the Inter-Ministerial Council
	2/17/2014	Establishment of Fukushima Council on Decommissioning and Contaminated Water
	8/18/2014	Decommissioning support is added to the Nuclear Damage Compensation Facilitation Corporation, which is reorganized as the Nuclear Damage Compensation and Decommissioning Facilitation Corporation
Operation of Decommissioning System (2014-present)	6/12/2015	3rd revision of the Mid-and-Long-Term Roadmap Amendment to NDF law enacted
	5/10/2017	Establishment of a reserve fund system to manage funds related to decommissioning
	9/26/2017	4th revision of the Mid-and-Long-Term Roadmap
	12/27/2019	5th revision of the Mid-and-Long-Term Roadmap

Table 14.2 shows the four principles which have been set forth in the Mid-and-Long-Term Roadmap's fifth revision, made on December 27, 2019. While Principles 2, 3, and 4 are exactly the same as those laid out in the previous revision of the Mid-and-Long-Term Roadmap on September 26, 2017 (the fourth revision), attention should be paid to the "coexistence of reconstruction and decommissioning", which has been rewritten as Principle 1.

Principle 1 in the 2017 version of the Mid-and-Long-Term Roadmap was simply stated to be '[s]ystematical reduction of risk in consideration of site conditions, efficiency, timeliness and proof while placing top priority on the safety of local citizens, the environment, and workers'. In the fifth revision, made on December 27, 2019, new phrasing was added: 'under the concept of "coexistence of reconstruction and decommissioning". … There has been gradual progress in the return of residents and reconstruction'.

The newly added 'coexistence of reconstruction and decommissioning' was likely due to the state of progress of decontamination work and partial lifting of evacuation orders in Specified Reconstruction and Revitalization Bases – areas in the vicinity of 1F, such as Namie, Futaba, and Okuma Town (please refer to Figure 1.1 map in Chapter 1). The phrase was written into the beginning of Principle 1 as a condition of consideration for the decommissioning work at 1F so that it would not hinder the progress of reconstruction in the surrounding area (Matsuoka, 2020a).

Table 14.2 Principles of the Mid-and-Long-Term Roadmap (2019)

Principle 1	Reduce risks systematically, under the concept of 'coexistence of reconstruction and decommissioning', with consideration for the site conditions, rationality, promptness, and certainty while placing top priority on the safety of locals, the surrounding environment, and workers. There has been gradual progress in the return of residents and reconstruction.
Principle 2	Move forward with mid- and long-term efforts while gaining the understanding of local and national citizens by actively releasing information and by thorough interactive communications with transparency.
Principle 3	Continuously update the Mid-and-Long-Term Roadmap in consideration of the site conditions, progress in the decommissioning and contaminated water management efforts, and the latest research and development (R&D) results.
Principle 4	Promote concerted efforts of Tokyo Electric Power Company, Nuclear Damage Compensation and Decommissioning Facilitation Corporation, and R&D institutions, the government of Japan, and other relevant organizations to achieve the goals indicated in this Mid-and-Long-Term Roadmap. The government of Japan should take the initiative in promoting the efforts to implement decommissioning safely and steadily.

Source: IMCCWD, Mid-and-Long-Term Roadmap towards the Decommissioning of TEPCO's Fukushima Daiichi Nuclear Power Station (2019), p. 2.

However, as we have just passed eleven years of Fukushima reconstruction in 2022, why has the issue of the coexistence of reconstruction and decommissioning appeared in the Mid-and-Long-Term Roadmap for the first time? Naturally, the dual problems of the reconstruction of Fukushima and decommissioning of 1F have existed in Fukushima since 2011. So why did they have to show up on the Mid-and-Long-Term Roadmap *now*?

How should we think about the relationship between Fukushima's reconstruction from a disaster and the decommissioning of 1F to begin with?

The Mid-and-Long-Term Roadmap was first published on December 21, 2011, and has had repeated revisions: first (July 30, 2012), second (June 27, 2013), third (June 12, 2015), fourth (September 26, 2017), and its fifth and final revision on December 27, 2019 (see Table 14.1).

In the first three versions of the Mid-and-Long-Term Roadmap (first edition, first and second revisions), the focus on accident handling was strong. Upon this premise, as objectives for the Mid-and-Long-Term Roadmap, the following two steps were emphasized: Step 1: achieving a steady downward trend in radiation levels and Step 2: achieving a controlled release of radioactive material and a significant reduction in radiation doses (first edition, p. 1). Moreover, a transition to ensure reliable stabilization of the nuclear plant as well as an 'Approach to Categorization in the Mid- and Long-Term' (first edition, p. 5) were shown as follows:

Phase 1: From the completion of Step 2 to the start of fuel removal from the spent nuclear fuel pool (target is within 2 years)

Phase 2: From the end of Phase 1 to the start of fuel debris removal (target is within ten years)

Phase 3: From the end of Phase 2 to the end of decommissioning (target is 30–40 years)

Although there were some changes in how it is worded, this framework continued from the first edition through the first and second revisions.

Table 14.3 shows the milestones of the Mid-and-Long-Term Roadmap (2019). This milestone table was first created for the third revision, in which it came to be organized into four milestone categories: (1) contaminated water management, (2) fuel removal from spent nuclear fuel pools, (3) fuel debris retrieval, and (4) radioactive waste management.

Table 14.4 shows transitions between milestones in the Mid-and-Long-Term Roadmap. The only measures against contaminated water shown in the milestone table are the control of the amount of contaminated water generated, and the treatment of stagnant water inside the reactor buildings. There is no mention of final disposal of the 1,293,044 of treated water stored in 1,061 tanks (according to the TEPCO water treatment portal site, as of March 3, 2022, treated water is referred to as 'tritiated water', but some collections contain other radioactive substances, such as radioactive caesium and strontium, which exceed emission standards).

In addition, in respect to the removal of nuclear fuel from the spent nuclear fuel pools (392 pools: 292 spent nuclear fuels, 100 new nuclear fuels), significant postponements in the schedule for the start of removal from Unit 1's spent nuclear fuel pool have been repeated with each revision of the Mid-and-Long-Term Roadmap, from 2020 (3rd revision), to 2023 (4th revision), to between 2027 and 2028 (5th revision). Similarly, the planned schedule for the start of fuel removal from Unit 2's spent nuclear fuel pools (615 pools: 587 spent nuclear fuels, 28 new nuclear fuels) has also seen repeated postponements, from 2020 (3rd revision) to 2023 (4th revision) to between 2024 and 2026 (5th revision).

Furthermore, due to delays in robotics development in the UK following the COVID-19 pandemic, the schedule for the start of removing fuel debris from Units 1, 2, and 3, which caused a meltdown of an estimated 800 to 900 tons, was reported to have been postponed from 2021 to 2022 or later (Nikkei, December 23, 2020).

Table 14.3 Mid-and-Long-Term Roadmap (2019) Milestones

Area	Description	Timing
1. Contaminated water management		
Contaminated water generation	Reduce to about 150m³/ day	Within 2020
	Reduce to about 100m³/ day or less	Within 2025
Completion of stagnant water removal/treatment	Complete stagnant water removal/ treatment in buildings•*	Within 2020
	Reduce the amount of stagnant water in reactor	FY 2022–FY 2024
	building to about a half of that in the end of 2020	
2. Fuel removal from spent fuel pools		
Complete fuel removal from Unit 1–6		Within 2031
Complete installation of the large cover at Unit 1		Around FY 2023
Start fuel removal from Unit 1		FY 2027– FY 2028
Start fuel removal from Unit 2		FY 2024–FY 2026
3. Fuel debris retrieval		
Start fuel debris retrieval from the first implementing Unit {Start from Unit 2, expand the scale of retrieval gradually)		Within 2021
4. Waste management		
Technical prospects concerning and the processing/ disposal policies and their safety		Around FY 2021
Eliminate temporary storage areas outside for rubble and other waste.**		Within FY 2028

Source: Mid-and-Long-Term Roadmap (2019), p. 13.

Note: FY = fiscal year.

*   Excluding the reactor buildings of Units 1 through 3, process main building, and high-temperature incinerator building.

** Excluding the secondary waste from water treatment and waste to be reused.

As for why the 'debris removal' had been solely implemented as the way to process radioactive waste, initially there had been discussions about alternatively using the 'water tomb' method or Chernobyl's 'sarcophagus method', but due to damage to Unit 2 and a request from Fukushima Prefecture governor for the 1F site as a vacant plot ('greenfield'), it has shifted to 'debris removal'. The first Mid-and-Long-Term Roadmap (created on December 21, 2011) clearly states that 'fuel debris removal will begin within 10 years after the accident'. This initial Mid-and-Long-Term Roadmap was drafted by TEPCO, the Agency for Natural Resources and Energy, and the Nuclear and Industrial Safety Agency, according to instructions from the Minister of Economy Trade and Industry (Minister Edano) and the minister in charge of Nuclear Accident Settlement and Prevention (Minister Hosono) on November 9, 2011. The road map was finalized at the government and TEPCO's Mid-to-Long-Term Countermeasure Meeting on 12 December 2011.

Based on this sequence of events, the government and TEPCO had already decided to use 'debris removal' as a basic strategy by the autumn of 2011. After that, they demonstrated a strong path dependency —which is a characteristic of Japanese policies and systems, through the idea that once something is decided, it does not change (Ministry of Economy, Trade and Industry, Committee for Reforming TEPCO and Overcoming 1F Challenges (TEPCO Committee), 2016). Until the latest version of the Mid-and-Long-Term Roadmap (5th revision, December 27, 2019), the policy had been consistent with using 'debris removal' as the way to process radioactive waste.

However, the question of how to deal with radioactive waste (including debris that has been removed), which increases in quantity as decommissioning work progresses, is the most important issue when considering the future of the 1F decommissioning (including the possibility of an 'interim end state'). It is necessary to focus on measures against radioactive waste in carrying out decommissioning work for 1F. From this perspective, we may need to stop and re-examine whether the strategy of debris removal is rational.

In any case, the target periods shown in the milestone table range from 2020 to 2028, but the parameters of the milestones themselves are extremely vague. The Mid-and-Long-Term Roadmap *should* be a strategic plan that is the foundation of 1F decommissioning policy, but in reality, it focuses on '1. Contaminated water management' and '2. Fuel removal from spent nuclear fuel pools', and these two goals are also uncertain. I have to say that at the moment, the road map still has strong characteristics of an 'accident management' plan.

The original Mid-and-Long-Term Roadmap shows the 'interim end state' of the mid- to long-term decommissioning process and should be a plan which functions as the basis for considering a long-term 'end state', but the current Mid-and-Long-Term Roadmap is not such a document.

In the following section, we consider the biggest concern of the current 1F decommissioning policy – the issue of management and disposal of treated radioactive water – and the relationship between expert and local knowledge, as well as think about what the lessons to be learned from Fukushima and what constitutes Fukushima reconstruction knowledge.

Table 14.4 Changes in Milestones in Revisions of the Mid-and-Long-Term Roadmap.

	3rd Revision (June 12, 2015)	*4th Revision* (September 26, 2017)	*5th Revision* (December 27, 2019)
① Contaminated water management	• Promotion of contaminated water management under three basic policies (removing the source of pollution, isolating water from the source of pollution, and prevent leakage of contaminated water) Aim to complete treatment of stagnant water by the end of 2020	• Continued promotion of contaminated water management in accordance with basic policy Reduction of contaminated water generation to about 150 m³/day • Aim to complete treatment of stagnant water by the end of 2020	• Continued promotion of contaminated water management in accordance with basic policy • Reduction of contaminated water generation to about 150 m³/day • Reduction of contaminated water generation to about 100 m³/day by the end of 2025 • Aim to complete treatment of stagnant water between 2022 and 2024
② Fuel removal from spent fuel pools	Unit 1: Start fuel removal in 2020 Unit 2: Start fuel removal in 2020 Unit 3: Start fuel removal in 2017 Unit 4: Completed fuel removal in December 2014 Units 5 & 6: Proper storage of spent fuel in pools for the time being	Unit 1: Start fuel removal in 2023 Unit 2: Start fuel removal in 2023 Unit 3: Start fuel removal mid-2018 Unit 4: Completed fuel removal in December 2014 Units 5 & 6: Proper storage of spent fuel in pools for the time being	Unit 1: Complete installation of a large cover by the end of 2023. Start fuel removal between 2027 and 2028 Unit 2: Start fuel removal between 2024 and 2026 Units 1–6: Aim to complete fuel removal by the end of 2031
③ Fuel debris retrieval	• Fuel debris removal policy to be decided 2 years later (2017) for each unit • Confirm method for removal by the first half of 2018 • Start fuel debris removal in first unit during 2021	• Confirm method of fuel debris removal for the first unit in 2019 • Start fuel debris removal for the first unit in 2021	• The first unit is Unit 2 • Start fuel debris removal for Unit 2 in 2021 and gradually increase scale of retrieval

(Continued)

Table 14.4 (Continued)

	3rd Revision (June 12, 2015)	4th Revision (September 26, 2017)	5th Revision (December 27, 2019)
④ Waste management	• Establish basic concept of processing/ disposal for solid radioactive waste in 2017	• Establish technical prospects concerning processing/ disposal policies and their safety in 2021	• Establish technical prospects concerning processing/ disposal policies and their safety in 2021 • Eliminate temporary storage areas outside for rubble and other waste in 2028

Source: Created from various revisions of the Mid-and-Long-Term Roadmap.

The Necessity of Collaboration between Expert and Local Knowledge in the 1F Decommissioning Policy

(a) The Subcommittee on Treated Water: The Mid-and-Long-Term Roadmap for 1F decommissioning policy shows an 'interim end state' for the original decommissioning process. The biggest factor in it not being a planning document serving as a basis for considering a long-term 'end state' is that it does not include an ultimate policy for the management and disposal of contaminated and treated water, which is still being generated through decommissioning tasks at a rate of 150 tons per day.

In the following, we focus on the activities of the Tritiated Water Task Force and its successor, the Subcommittee on Handling of the ALPS Treated Water (hereinafter referred to as the Subcommittee on Treated Water), from the perspective of forming 'socially robust knowledge' and the 'democratization of knowledge', suggested by research of science-policy interfaces, which deals with the relationship between risk expertise in science and technology (scientific and technological expertise as well as cultural and social expertise) and the government and citizens (Jasanoff, 2003, Spruijt, 2014).

The activities of the Subcommittee on Treated Water have attempted to form 'socially robust knowledge' through policy dialogue between groups of experts from various fields and local residents as well as a diverse group of citizens. This report targets the activities of the Subcommittee on Treated Water, considers the relationship between expert and local knowledge in 1F decommissioning policy and the reconstruction of Fukushima, as well as the relationship between 'socially robust knowledge' and the 'democratization of knowledge', and clarifies the matter of Fukushima reconstruction knowledge.

The Tritiated Water Task Force, the predecessor of the Subcommittee on Treated Water, was established on December 25, 2013. The Tritiated Water

Task Force published the Tritiated Water Task Force Report (hereinafter referred to as 'the Report') on June 3, 2016, ending its role (Ministry of Economy, Trade and Industry, 2013, 2014 & 2016).

In the Report, it states that a technical assessment was conducted on a variety of options to provide basic data for deciding a long-term policy for handling water treated with multi-nuclide removal equipment (tritiated water). In addition, a note was added to the Report that opinions would not be coordinated among stakeholders and that options would not be consolidated (Tritiated Water Task Force Report, p. 13). The Report states that a technical assessment was conducted in which cases for assessment were set based on equivalent conditions of management for 11 policy options that combined five methods of disposal (geosphere injection, offshore release, vapour releases, hydrogen release, underground burial) and pretreatment. Table 14.5 shows the results of the technical assessment, divided by the time (in months) and cost (in billions of yen) required to complete disposal.

The Tritiated Water Task Force's Report states, 'it is hoped that future discussions about handling tritiated water will be advanced in a comprehensive manner, touching upon both technical perspectives, such as feasibility, economic efficiency, and duration, as well as social perspectives, such as damage caused by harmful rumor.' In response to this point, on November 11th, 2016, the Subcommittee on Treated Water was established by and under its parent committee, the Committee on Countermeasures for Contaminated Water Treatment.

The Subcommittee on Treated Water met twice in 2016, four times in 2017, and six times in 2018. Of particular note was the Explanation & Public Hearing on the Handling of the ALPS Treated Water (hereinafter referred to as 'the Explanation and Public Hearing') held on August 30–31, 2018, in Tomioka (Fukushima Prefecture), Kōriyama (Fukushima Prefecture), and Chiyoda (Tokyo).

A total of 44 people expressed their opinions at the three venues (14 in Tomioka Town, 14 in Kōriyama City, and 16 in Tokyo), out of a total of 274 participants (101 in Tomioka, 88 in Kōriyama, and 85 in Tokyo). At the same time, written opinions were also solicited, and 135 people submitted theirs over a period of 39 days (Ministry of Economy, Trade and Industry Secretariat of the Subcommittee on Handling of the ALPS Treated Water, 2018a, p. 2).

Table 14.5 Options for Handling Tritiated Water and Technical Assessment

Disposal Method	Time Required to Complete Disposal	Disposal Cost
Geosphere injection	69–156 months	17.7–397.6 billion yen
Offshore release	52–88 months	1.7–3.4 billion yen
Vapour release	75–115 months	22.7–34.9 billion yen
Hydrogen release	68–101 months	60–100 billion yen
Underground burial	62–98 months	74.5–253.3 billion yen

Source: Tritiated Water Task Force Report, Ministry of Economy, Trade and Industry (2016).

At the Explanation and Public Hearings at each of the three venues, there was a great deal of concern with and opposition to the idea that the policy of offshore release of treated water suggested by the Tritiated Water Task Force's Report was the most efficient policy in terms of time and cost. Regarding the other four policy options shown in the Report besides offshore release (geosphere injection, vapour release, hydrogen release, underground burial), concerns about adverse environmental effects (from vapour or hydrogen release) and difficulty in monitoring (of geosphere injection or underground burial) were also expressed. A proposed policy alternative to the five policy options in the report was long-term above-ground storage in large tanks, such as those used for storing oil.

The author was in attendance at the Tokyo venue on the afternoon of August 31, 2018, where activists including anti-nuclear citizen's groups participated in large numbers, including those who expressed their opinions. Presenters of opinions expressed them unilaterally, and the atmosphere was very noisy, with jeers and angry roars flying back and forth. In a deliberation process, it is important for participants to have a constructive discussion, confirm the rationales and differences of each claim, and encourage the discovery of new approaches to the task at hand. The formation of that kind of forum for discussion is a prerequisite for the formation of 'socially robust knowledge' and the 'democratization of knowledge' needed to do so, as is suggested by science–policy interface research (Jasanoff, 2003).

As its name implies, the aim of the explanation and public hearing in August 2018 seems to have been to break away from conventional hearing method of just listening to the opinions of selected residents, and to have two-way communication between the explanation given to citizens by the Subcommittee on Treated Water (expert knowledge) and the expression of opinions by citizens (local knowledge). We could praise this as a very ambitious effort, but it must be said that the result was terribly inadequate.

In April 13, 2021, the government of Japan officially decided that the 1F-treated water would be released offshore in 2023, despite fishermen's strong voice against it.

(b) Collaboration between Expert and Local Knowledge – Challenges Facing Fukushima Reconstruction Knowledge: After digging deeper into the formation of 'socially robust knowledge' and the 'democratization of knowledge' suggested by research of science-policy interfaces, the two-step approach from the Tritiated Water Task Force to the Subcommittee on Treated Water was adopted, and after narrowing down the policy options in two stages, a third step was implemented in attempting to interact with the community. It should be pointed out that this was the biggest issue.

The government (via the Ministry of Economy, Trade and Industry [METI]), was the manager of the situation, and as a first step had envisioned that the Tritiated Water Task Force would present policy options from a technical perspective. As the second step, the Subcommittee on Treated Water would narrow down the policy options in consideration of social aspects, such as the damage caused by harmful rumours about the options, and as the third step, it would have a dialogue with citizens and local residents. Such a scenario may have led to decisions on policy from the government.

The policymakers seemed to regard this sort of three-step approach as being able to ensure the legitimacy of the 1F decommissioning policy (Ministry of Economy, Trade and Industry Secretariat of the Subcommittee on Handling of the ALPS Treated Water, 2018b and 2019). Through the policy for the disposal of treated water with extremely high levels of uncertainty and complexity, the legitimacy of this process would be its cornerstone. However, in order to properly carry out the process of creating policy options as a 'knowledge democratization' process to form 'socially robust knowledge', it may have been necessary to design the three aspects of technical analysis, social analysis, and dialogue with the community into one process simultaneously, rather than dividing it into three sequential stages. This would prevent the over-privileging of the technical account.

In terms of options for disposing of treated water, it is also very puzzling that the option of tank storage was excluded from consideration by the Tritiated Water Task Force from the outset. In the debate over the management and disposal of high-level radioactive waste (HLW), long-term aboveground storage is based on the principle of reversibility in guaranteeing powers of policymaking for future generations. This is a well-known policy option (Matsuoka, 2019), and for tritium, which has a much shorter half-life (12 years) than HLW (2.14 million years for neptunium-237, 1.53 million years for zirconium-93), ground storage is a sufficiently realistic policy option.

In any case, when it comes to the activities of the Subcommittee on Treated Water, based on this evaluation of policy options by the Tritiated Water Task Force (whose reliability is extremely unclear), I am compelled to say that the preconditions themselves had collapsed. Additionally, it goes without saying that the explanation and public hearing (dialogue with the local community) held by the Subcommittee on Treated Water based on its report did not create 'socially robust knowledge'.

The tragedy of the Subcommittee on Treated Water, which had held discussions, as well as an explanation and public hearing based on a report with questionable reliability, seems to have been caused by the separation of technical analysis (by the Tritiated Water Task Force) and social analysis (by the Subcommittee on Treated Water) of policy options from the start.

As lessons of science–policy interfaces study suggests, the formation of 'socially robust knowledge' and the 'democratization of knowledge' required that the technical and social aspects of the task at hand should not be separated. Only through the collaboration of diverse technical and social experts, and further collaboration between these diverse experts and a variety of non-specialized knowledge possessed by civil society, is it possible to form 'socially robust knowledge', which equates to 'Fukushima reconstruction knowledge', or in other words, lessons learned from Fukushima.

Work Needed to Make the Coexistence of Reconstruction and Decommissioning Possible

To facilitate the coexistence of reconstruction and decommissioning in Fukushima, it is first necessary to clarify 1F decommissioning governance and to provide the local community with options for interim states of 1F decommissioning. Second, it is important, under clear 1F decommissioning governance, to create mechanisms for dialogue between those performing 1F decommissioning work and the local community and for making the 1F decommissioning process itself a local resource (1F Decommissioning End State Research Meeting, 2020).

(a) 1F Decommissioning Governance and Interim States: It is important to create clear 1F decommissioning governance. 1F decommissioning policy depends on so many institutions such as METI, TEPCO, NDF (Nuclear Damage Compensation and Decommissioning Facilitation Corporation), and the Japan Atomic Energy Agency. The present situation does not clarify who will be wholly responsible for proceeding with the decommissioning of 1F. To do that, establishing the 1F decommissioning governance system is essential. This governance system should be clarified in terms of who is going to consider, discuss, decide on, implement, evaluate, and revise decommissioning policy. We would do well to recall the famous words of business scholar Alfred Chandler: 'structure follows strategy' (Chandler, 1969).

Another important point is how we take science–policy interfaces in 1F decommissioning policy into account (Matsuoka, 2019). Which experts, which citizens and local residents form what sort of 'place' for careful deliberation? Who is the leader of that 'place'? What are the agenda (framing), rules, process design, micro–macro loops? These are also important points. When considering the 'democratization of knowledge' and the creation of 'socially robust knowledge', it is important not to separate technical and social aspects.

Furthermore, it is difficult to indicate a final end state at this point in 1F decommissioning, and it is necessary to consider a 'provisional end state' for decommissioning, or 'interim state'. Policies and plans with unclear exit strategies and goals only lead to public anxiety and distrust. To that end as well, it is important to first determine an interim state with a well-defined timeline for 20 years (2031) and 30 years (2041) after the accident and to proceed with 1F decommissioning work.

One plan for interim states of 1F decommissioning is to think of it in two stages: Interim State 1 – stable long-term storage of waste other than debris and Interim State 2 – stable long-term storage of high-level radioactive waste such as debris (through use of storage tanks, etc.).

In addition, regarding long-term storage of radioactive waste generated by 1F decommissioning work and the establishment of an interim state, it is essential to plan for both on- and off-site in an integrated manner. It should be noted that the 1F site is 348.5 hectares, and the interim storage facility that surrounds it, which manages and treats decontamination waste, covers 1,600 hectares (as of March 2020, acquisition of land includes about 1,200 hectares). The interim storage facility has a 30-year deadline, set from the start of delivery of decontamination waste in 2015; the function of the facility is set to be completed by 2045. The use of the land of the interim storage facility site (which is in contact with the area where further decommissioning of 1F and reconstruction are progressing) over the next 23 years and beyond should be integrally considered as a continuous space.

Furthermore, in order to pass on lessons learned from this nuclear disaster to future generations, it is necessary to preserve 1F as a legacy of the accident to the extent that is possible.

(b) Making the 1F Decommissioning Process a Local Resource and the Concept of Arts and Sciences: The key to producing a coexistence between reconstruction and decommissioning in Fukushima is converting the 1F decommissioning process itself into a local resource and creating a 'virtuous cycle' mechanism between the decommissioning and reconstruction processes. We should not think of the 1F decommissioning process only through the narrow scope of local companies' participation in decommissioning work or of decommissioning tourism or dark tourism.

In order to progressively pass on the lessons learned from the severe Level 7 incident at the Fukushima Daiichi nuclear power plant, a 20th-century system of science and technology, for future generations, it will be necessary to overcome the harmful effects of such systems which have lost their completeness due to piecemeal engineering. To do that, it is important to consider the perspective of integrating art and science, as exemplified by Leonardo da Vinci during the Renaissance period 500 years ago. In learning the lessons of the history of the 1F accident, we can consider the reconstruction of a collaborative relationship between the arts, which find value in the diversity of humans and human life and express their wholeness, and the systems of science and technology.

As a place to break through the limits of 20th-century systems of science and technology and rebuild a new, collaborative relationship between cultural arts and science and technology in the 21st century, the Hamadori area in Fukushima Prefecture, which centres on 1F, is the most suitable place in the world. A knowledge centre for arts and academia is expected to be formed in Fukushima. This will connect science and art to create local resources from the 1F decommissioning process that will form a new collaborative relationship between cultural arts (assuming interdisciplinary academic research on sustainability and resilience) and science and technology (assuming a wide range of cultural arts including architecture, design, and literature). This process will facilitate the development of human resources (nurturing the Leonardo da Vinci of the 21st century) for the formation of disaster-resistant and sustainable communities.

In taking advantage of the collaborative relationship between art and science, creating new cultural industries and innovations in science and technology, and demonstrating models of creative reconstruction from the COVID-19 pandemic (Build Back Better, green recovery (Toyoda, 2020)), the coexistence of reconstruction and decommissioning in Fukushima can be considered to be 'Fukushima reconstruction knowledge' – and our lesson learned from Fukushima.

References

Chandler Alfred, D. (1969). *Strategy and Structure: Chapters in the History of the American Industrial Enterprise.* London: The MIT Press; Yuko Aruga (translator). 組織は戦略に従う. Diamond Inc., 2004.

1F Decommissioning End State Research Meeting (2020). *1F 廃炉の先研究会・中間報* [Interim report from the 1F decommissioning end state research meeting]. Waseda University Fukushima Hirono Future Creation Research Center, May 6th, 2020.

IMCCWD; Inter-Ministerial Council for Contaminated Water and Decommissioning Issues, Government of Japan. (2019). *Mid-and-Long-Term Roadmap towards the Decommissioning of TEPCO's Fukushima Daiichi Nuclear Power Station.* https://www.meti.go.jp/english/earthquake/nuclear/decommissioning/pdf/20191227_3.pdf (accessed 25 February 2022)

Jasanoff, Sheila. (2003), Breaking the waves in science studies. *Social Studies of Science, 33* (2), pp. 389–400.

Matsuoka, Shunji. Iwaki OtentoSUN Enterprise Cooperative (editor) (2013). フクシマから日本の未来を創る：復興のための新しい発想 [Creating the Future of Japan from Fukushima: New Ideas for Reconstruction]. Tokyo: Waseda University Press.

Matsuoka, Shunji (2017).『フクシマの教訓』と早稲田大学ふくしま広野未来来創造リサーチセンターの挑戦． アトモス(日本原子力学会誌) ["Lessons from Fukushima" and the challenges of Waseda University's Fukushima Hirono Future Creation Research Center.] *Atmos (Journal of the Atomic Energy Society of Japan), 59* (9), pp. 2–3.

Matsuoka, Shunji (2018). 社会イノベーションと地域の持続性：場の形成と社会的受容性の醸成 [Social Innovation and Regional Sustainability: Forming Places & Fostering Social Acceptability]. Tokyo: Yuhikaku Publishing.

Matsuoka, Shunji (2019). 原子力災害からの地域再生と 1F 廃炉政策：福島復興知を考える. [Regional revitalization from nuclear disaster & 1F decommissioning policy: Considering Fukushima reconstruction knowledge.] *Journal of Environmental Information Science, 48* (4), pp. 40–48.

Matsuoka, Shunji (2020a).『復興と廃炉の両立』を考える：東日本大震災と福島復興. [Considering the coexistence of reconstruction and decommissioning: The Great East Japan Earthquake and the reconstruction of Fukushima.] *Journal of Asia-Pacific Studies (Bulletin of the Waseda University Graduate School of Asia-Pacific Studies), 40*, pp. 27–43.

Matsuoka, Shunji (2020b). ポスト・トランス・サイエンスの時代における 専門家と市民：境界知作業者, 記録と集合的記憶, 歴史の教訓. [Experts and citizens in the era of post-transformer science: Creators of "boundary knowledge", records and collective memory, and lessons from history.] *Journal of Environmental Information Science, 49* (3), pp. 7–16.

Ministry of Economy, Trade and Industry (2013). トリチウム水タスクフォース規約(案) [Tritiated Water Task Force Regulations (draft)] (Tritiated Water Task Force, Meeting 1, Document 1). https://www.meti.go.jp/earthquake/nuclear/pdf/131225/131225_01c.pdf (accessed June 10th, 2019)

Ministry of Economy, Trade and Industry (2014). トリチウム水タスクフォース『これまでの議論の整理』(案) [Tritiated Water Task Force – Summary of Previous Discussions (draft)] (Tritiated Water Task Force, Meeting 8, Document 3). https://www.meti.go.jp/earthquake/nuclear/pdf/140424/140424_02_005.pdf (accessed June 10th, 2019)

Ministry of Economy, Trade and Industry (2016). Tritiated water task force report. Tritiated Water Task Force, June 2016 https://www.meti.go.jp/earthquake/nuclear/osensuitaisaku/committttee/tritium_tusk/pdf/160603_01.pdf (accessed June 10th, 2019)

Ministry of Economy, Trade and Industry Committee for Reforming TEPCO and Overcoming 1F Challenges (TEPCO Committee) (2016). 東京電力改革・1F問題委員会 提言原案骨子案 [Outline of draft proposal for the committee for reforming TEPCO and overcoming 1F challenges]. https://www.meti.go.jp/committee/kenkyukai/energy_environment/touden_1f/pdf/006_01_00.pdf (accessed June 10th, 2019)

Ministry of Economy, Trade and Industry Secretariat of the Subcommittee on Handling of the ALPS Treated Water (2018a). 説明・公聴会について [Regarding explanation & public hearing]. https://www.meti.go.jp/earthquake/nuclear/osensuitaisaku/committttee/takakusyu/pdf/010_02_00.pdf (accessed June 10th, 2019)

Ministry of Economy, Trade and Industry Secretariat of the Subcommittee on Handling of the ALPS Treated Water (2018b). 第10回小委員会議事録 [Minutes from the 10th Subcommittee Meeting]. https://www.meti.go.jp/earthquake/nuclear/osensuitaisaku/committttee/takakusyu/pdf/011_01_01.pdf (accessed June 10th, 2019)

Ministry of Economy, Trade and Industry Secretariat of the Subcommittee on Handling of the ALPS Treated Water (2019). 貯蔵継続及び処分方法について [Regarding methods for continued storage & disposal]. https://www.meti.go.jp/earthquake/nuclear/osensuitaisaku/committttee/takakusyu/pdf/013_04_01.pdf (accessed October 8th, 2019)

Murakami, Yoichiro (2005). *安全と安心の科学* [The Science of Safety & Peace of Mind]. Tokyo: Shueisha Shinsho.

Spruijt, P. et al. (2014). Roles of scientists as policy advisers on complex issues: A literature review. *Environmental Science and Policy*, *40*, pp. 16–25.

Toyoda, Toshihisa (2020). The framework of international cooperation for disaster risk reduction – A reconsideration with special reference to "build back better". *Journal of International Cooperation Studies*, *27* (2), pp. 1–15.

15 The Long-term Impact of Disasters and Looking Forward

Claire Leppold, Alison Lloyd Williams,
Akihiko Ozaki and Sudeepa Abeysinghe

As the world currently reels from the COVID-19 pandemic and the increasing urgency of climate change, it is all the more vital to reflect on what we can learn about recovery from previous crises such as 3.11. This collection has shown how, more than ten years on from the disaster, different concerns and priorities have emerged for the communities that continue to live with its legacy, both in Fukushima Prefecture and beyond. These experiences have critical relevance for us today in understanding processes of disaster recovery and how to live long-term with disaster and disaster risk.

This book has brought together perspectives from different disciplines and expertise, including those which reflect first-hand experience of the 3.11 disaster. It includes chapters from those who have worked as practitioners in health care (Onoda and Sato, Chapter 2; Hori, Chapter 3), education (Takamura, Chapter 4) and law (Mimura, Chapter 5) and those who have approached the disaster and post-disaster recovery through academic research (Chapters 6–14). There are also chapters where authors directly reflect their dual roles as both volunteers and researchers after the disaster (Lee, Chapter 6; De Togni, Chapter 12) and as both residents and professionals in Fukushima (Onoda and Sato, Chapter 2; Hori, Chapter 3; Takamura, Chapter 4).

The diversity of perspectives from each author helps us to see the disaster from different viewpoints and to better understand the different ways that 3.11 played out and was experienced by different communities. Through presenting this diversity of perspectives, this book emphasises the sentiment that there is no single or definitive version of the disaster or its impacts. There is a need for a pluralistic and holistic understanding of the experience of disaster and the disaster recovery.

Disaster Recovery – Where Have We Been, and Where Are we Going?

Disaster recovery has been referred to as the 'least understood' part of the disaster cycle (Smith and Wegner, 2007). The immediate impacts and responses to disasters have historically gained more attention, leaving recovery to the wayside. However, this trend is changing, with increasing emphasis placed on the need to better understand the long-term recovery process

DOI: 10.4324/9781003182665-19

(Tierney and Oliver-Smith, 2012). There is a growing body of work now focusing on the experiences of people and communities in the months and years following a disaster (Gibbs et al., 2013) and the policy process needed to address the long-term challenges of disaster recovery and address inequities in this process (Finucane et al., 2020). New, practical resources have emerged for those involved in supporting disaster recovery (Recovery Capitals, 2021). Still, there remains a need for further research. In the context in which this book is published, evidence on disaster recovery is more relevant than ever before, with many (national and local) governments, public health institutions and communities looking for pathways to societal recovery from the COVID-19 pandemic (Few et al., 2020).

In this book, we set out to harness critical public health and social science perspectives to extend the literature on health, wellbeing and community recovery after disaster. Each chapter has presented different perspectives on the ways that social structures, institutions, relationships and experiences have underpinned the 3.11 disaster and the process of recovery. In Chapter 1, we situated the case of the 3.11 disaster within the wider field of disaster studies. In presenting the 3.11 disaster and existing evidence on health and wellbeing, we outlined the complex nature of literature on disaster recovery and centred the focus of the book on understanding social aspects of recovery, health and wellbeing.

Chapters 2 through 5 explored 'Reflections from the Field', drawing on the perspectives of practitioners in Fukushima. In Chapter 2, Onoda and Sato reflected on their experiences as nurses in the ten years since the disaster. The authors looked back at the immediate decisions they made in their professional and personal lives, as well as their continued work together to both rebuild their nursing department and inspire hope for the next ten years. Chapter 3 comes from another health care professional, Hori, a psychiatrist who discussed approaches to supporting mental health after the disaster. In Chapter 4, Takamura, a schoolteacher, spoke to the various activities that have been set up with and for young people to promote learning and empowerment during recovery from a nuclear disaster. In Chapter 5, Mimura wrote as a lawyer about his experience with legal issues, including disaster-related deaths, and working to assist Fukushima Prefecture residents (or former residents) in the process of navigating the complex legal system for compensation. These practitioner chapters provide a window into on-the-ground experiences and professional expertise in the long-term aftermath of the disaster, through reflecting on how the disaster impacted the professional practices of health, the law and education and how the disaster context impacted on these authors' own practices and orientations towards their professional roles.

In Part II, 'Living with Risk', Elstow (Chapter 6), Murakami (Chapter 7) and Epstein (Chapter 8) each present perspectives on navigating different types of risk following the disaster. Elstow explored the meaning of radiation monitoring and measuring and argues that these actions go beyond an overt assessment of implications for human health to also become a form of meaning-making for people and communities. Murakami similarly suggested that the risk measurements that see uptake will depend on what is meaningful to each community or society; however, he also suggests that there are non-radiological measurements of health and wellbeing that are relevant in Fukushima but potentially left out when the dominant focus goes to radiation. Finally, Epstein drew on ethnography to look at another way of categorising risk measurements – in this case relating to mental health risks – in order to inform models of care and support delivered to disaster-affected communities. From radiation monitoring practices by community members to the academic practice of determining which risk indicators to use in research to the ways that mental health intervention practices are built on certain measurements, each of these chapters gives new insight into what it means to live with risk in the long-term aftermath of the 3.11 disaster. These chapters also all speak to the juxtaposition between, on one hand, the social and professional quantification of risk and, on the other, the social experience of risk. This highlights the tensions inherent in these two ways of understanding and experience. This draws through themes from Mimura (Chapter 5) and Hori (Chapter 3), where the reification of risk and the impacts of living with risk – there, through the process of conferring a diagnosis of mental illness or through the process of seeking compensation – have profound implications on individual processes of recovery.

Part III, 'Social Difference and Inequality', examines these concepts through multiple lenses in relation to the 3.11 disaster. In Chapter 9, Lee analysed gender issues after 3.11 through a review of gender and disaster in Japan and cases studies of ways that women, often in challenging circumstances, have taken instrumental roles to support their communities through recovery from the 3.11 disaster. In Chapter 10, Uekusa and Lee examined pre-existing inequities and socially produced vulnerabilities faced by 'marriage-migrant' women, and their experiences in rural Tōhoku following the 3.11 disaster. In looking to recovery, both Chapters 9 and 10 emphasised the importance of addressing pre-existing (pre-disaster) inequities in order to promote fair outcomes in disaster recovery. Finally, in Chapter 11, Tanaka took a more macro-sociological approach to investigate the spread of stigma about Fukushima after the disaster and the impacts that it has had. These chapters contribute to the growing disaster literature that emphasises the way in which crises magnify and exacerbate existing social differences and social inequalities. Tanaka's work also demonstrates that disaster events can create new forms of social difference, for example here through processes of place-based stigma linked to the (perceived) effects of nuclear contamination.

Last, in Part IV, 'Community Engagement and Wellbeing', there is a strong focus on the strength and resilience of communities and what can be accomplished when people work together. In Chapter 12, De Togni explored the empowerment of self-evacuated single mothers who pursued their own rights and engaged in disaster-related litigation, through working together with volunteers from a non-governmental organisation. In Chapter 13, Lloyd Williams and Goto discussed the impact that children and young people can have in disaster recovery and presented a research study in which students in

Date City used participatory theatre to speak directly to their communities about their experiences and ideas for change. In Chapter 14, Matsuoka zoomed out to consider the Japan-level road map that has been constructed for recovery from the 3.11 disaster, the dual roles of reconstruction and decommissioning (of the nuclear plant) and the importance of integrating the arts and sciences for this purpose. These chapters link to ideas on recovery processes and empowerment in the long-term aftermath of the disaster presented by Takamura, in relation to the knowledge of young people (Chapter 4), and by Onoda and Sato on growth, professionally and personally, after the disaster (Chapter 2).

Together, these chapters give us a chance to think in new ways about how we experience and live with disaster and what that means for how we manage disasters in the future. They respond to the idea of recovery as the 'least understood' part of disasters by illustrating multiple examples of what has occurred in the recovery phase after the 3.11 disaster – not looking to recovery as a completed process but as one that is in progress, involving complex interwoven elements. This includes the interplay between the personal and the social, between physical and mental health, between lived experience and 'official' accounts of disaster, between policy and practice, between reflecting on the past and looking to the future and, crucially, what that means for life now, in the present, in communities affected by disaster.

In this way, the chapters in this book have moved beyond 'disaster medicine' (see Chapter 1) to explore not just the health impacts of 3.11 but more specifically how people live through, perceive and recover from this experience. By going beyond the immediate, physical 'response' to the emergency, the perspectives and approaches taken by authors in this collection help us to look again at how disasters like 3.11 are represented and experienced by engaging with issues related to longer-term experience and recovery.

Looking to the Future

In this collection, the authors have pointed to various ways that we can learn to 'live better' with disaster risk and become more resilient to future crises. This includes inviting those who are often marginalised from such discussions to join the conversation (Takamura, Chapter 4; Lloyd Williams and Goto, Chapter 13; Uekusa and Lee, Chapter 10), working to strengthen organisations that can facilitate constructive dialogue between people affected by policies and the authorities involved in policy-making to achieve a more inclusive process (De Togni, Chapter 12), drawing on lessons from disaster deaths to improve responses to future disasters (Mimura, Chapter 5), disrupting the spread of stigmatising narratives in media (Tanaka, Chapter 11) and bringing together the arts and sciences into a productive discussion to document and preserve lessons from disasters and ways forward (Matsuoka, Chapter 14; Lloyd Williams and Goto, Chapter 13). As noted earlier, this book is itself a space where different viewpoints and experiences have come together. The authors have very different backgrounds, experiences and priorities but, through this volume, have a common aim: to share learning and create conversations between different communities of knowledge. We hope that this can serve as a touchpoint for the rest of society to draw on as globally we continue to grapple with new and ongoing crises.

Between 2011 and 2015, the International Commission on Radiological Protection (ICRP) held regular 'dialogue meetings' in Fukushima Prefecture, bringing together members of local affected communities with international specialists in radiology to explore how to 'find ways to meet the daily challenges of long-term rehabilitation' (IRSN, 2020). From 2016 to 2020, the programme continued, shifting from L'Institut de Radioprotection et de Sûreté Nucléaire' (ISRN) organisation to being organised by local actors in Japan to continue to support returning evacuees. As we suggest above, dialogue remains a key way forward for community recovery and resilience building in Fukushima and beyond. While the ICRP meetings focused in the main on radiation protection, there was a growing recognition of the need to involve cross-community networks, as well as the importance of maintaining open lines of communication and knowledge exchange.

Creating space for dialogue remains critical in order for different forms of knowledge, insight and lived experience to speak to each other and cross-fertilise. Disasters are not simple occurrences that can be completely captured through any one perspective. As we discussed in the Introduction, 3.11 is often referred to as a 'triple' or 'compound' disaster. There is increasing recognition, highlighted by social scientists, that disasters are complex, lengthy and multidimensional. There is growing attention to the cascading effects that disasters can have and cases in which secondary effects of disasters can lead to even greater impacts than the primary event (Pescaroli and Alexander, 2016). The 3.11 disaster is often described in this way, beginning with an earthquake, which caused a tsunami, which led to a nuclear disaster. However, it could be argued that the 'cascades' do not end when the hazards themselves end. The social, political, environmental, cultural and human effects of disasters can continue cascading long after the physical event is over. This book begins to uncover what some of those cascades have looked like in Fukushima, bringing together different chapters to better understand the different dimensions of disaster, how they interact with each other and how they impact individuals, families and communities.

At the same time, it should be recognised that there are gaps remaining in the literature. Theoretically, the definition of where disasters begin and end can be murky, and some scholars emphasise that disasters should be thought of as processes rather than 'events' (Kelman, 2018). This book has not intended to determine the 'end' of the disaster or its recovery process in Fukushima or resolve process-versus-event debates. Second, while some chapters of this book focus specifically on the nuclear disaster, others look at the combined aftermath of all three disaster elements involved in 3.11. These varying (often implicit) approaches reflect the wider definitional issues present in the field of disaster studies, the thorough unpacking of which is beyond the scope of the present volume. There has been limited research to date on disaster recovery and health and wellbeing in contexts where complex, cascading disasters have occurred, and while this book begins to broach this topic, there is still a global need for further research. It is vital to try and better understand the different dimensions of disaster, how they interact with each other and how they impact individuals, families and communities in order that we can learn more about how to live with the complexity of disaster in the future.

This collection is unique in providing long-term perspectives on the 3.11 disaster, taking an interdisciplinary approach to link critical social science perspectives on health and wellbeing following a disaster and bringing together the perspectives of both academics and practitioners. It is our hope that the book will bring renewed attention to the social experience of health and disaster recovery and that it will serve as an impetus for further work on the long-term aftermath of disasters.

References

Few, R., White, C., Chhotray, V., Armijos, T., Tebboth, M., Shelton, C. and Forester, J., (2020) Covid-19 crisis: lessons for recovery. What can we learn from existing research on the long-term aspects of disaster risk and recovery? *The British Academy* [Online]. Available at: https://www.thebritishacademy.ac.uk/publications/covid-19-crisis-lessons-recovery/ [Accessed: 15 August 2021].

Finucane M.L., Acosta, J., Wicker, A. and Whipkey, K. (2020) Short-term solutions to long-term challenge: Rethinking disaster recovery planning to reduce vulnerabilities and inequities. *International Journal of Environmental Research and Public Health* Vol. 17 (2), p. 482.

Gibbs, L., Waters, E., Bryant, R.A., Pattison, P., Lusher, D., Harms, L., Richardson, J., MacDougal, C., Block, K., Snowdon, E., Gallagher, H.C., Sinnott, V., Ireton, G. and Forbes, D. (2013) Beyond bushfires: community, resilience and recovery – a longitudinal mixed method study of the medium to long term impacts of bushfires on mental health and social connectedness. *BMC Public Health*, Vol. 13, p.1036.

IRSN (2020) Kotoba: *Dialogues in Fukushima* [Online]. Available at: https://www.irsn.fr/EN/Kotoba-EN/Pages/Kotoba-EN_Introduction.aspx [Accessed: 17 August 2021].

Kelman I. (2018) Connecting theories of cascading disasters and disaster diplomacy. *International Journal of Disaster Risk Reduction*, Vol. 30, pp. 172–179. doi:10.1016/j.ijdrr.2018.01.024

Pescaroli G. and Alexander D. (2016) Critical infrastructure, panarchies and the vulnerability paths of cascading disasters. *Natural Hazards (Dordrecht)*, Vol. 82 (1), pp. 175–192. doi:10.1007/s11069-016-2186-3.

Recovery Capitals (ReCap) (2021) *Guide to Disaster Recovery Capitals* [Online]. Available at: https://recoverycapitals.org.au/ [Accessed: 15 August 2021].

Smith, G.P. and Wegner, D. (2007) Sustainable disaster recovery: operationalizing an existing agenda. In: Rodriguez, H., Quarantelli, E.L. and Dynes, R.R. (eds). *Handbook of disaster research*. New York City: Springer, pp. 234–257.

Tierney, K. and Oliver-Smith, A., (2012) Social dimensions of disaster recovery. *International Journal of Mass Emergencies and Disasters*, Vol. 30 (2), pp. 123–146.

Index

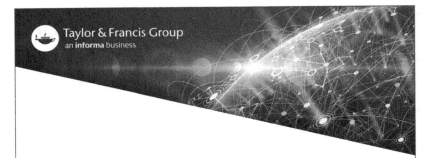

Printed in the United States
by Baker & Taylor Publisher Services